# Psychosocial Interventions with Physically Disabled Persons

# MIND AND MEDICINE

Series Editors
*Leonard S. Zegans, M.D.*
*Lydia Temoshok, Ph.D.*

Previous volumes in this series

*Emotions in Health and Illness
Theoretical and Research Foundations*
edited by Lydia Temoshok, Ph.D.,
Craig Van Dyke, M.D.,
and Leonard S. Zegans, M.D.

*Emotions in Health and Illness
Applications to Clinical Practice*
edited by Craig Van Dyke, M.D.,
Lydia Temoshok, Ph.D.,
and Leonard S. Zegans, M.D.

*Psychosocial Interventions with Sensorially Disabled Persons*
edited by Bruce W. Heller, Ph.D.,
Louis M. Flohr, M.D.,
and Leonard S. Zegans, M.D.

# Psychosocial Interventions with Physically Disabled Persons

*Editors*

Bruce W. Heller, Ph.D.

Louis M. Flohr, M.D.

Leonard S. Zegans, M.D.

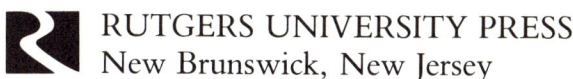
RUTGERS UNIVERSITY PRESS
New Brunswick, New Jersey

Copyright © 1989 by Rutgers, The State University

All rights reserved

Library of Congress Cataloging-in-Publication Data

Psychosocial interventions with physically disabled persons.

  (Mind and medicine)
  Companion v. to: Psychosocial interventions with sensorially disabled persons.
  Includes bibliographies and index.
ISBN 0-8135-1423-1 (cloth)  ISBN 0-8135-1424-X (pbk.)
  1. Physically handicapped—Psychology.  2. Physically handicapped—Rehabilitation—Social aspects.
I. Heller, Bruce W.  II. Flohr, Louis M.  III. Zegans, Leonard S.
IV. Psychosocial interventions with sensorially disabled persons.  V. Series.
[DNLM: 1. Handicapped—psychology.  2. Psychology, Social.
  3. Rehabilitation.  HV 3011 P974]
RC451.4.H35P76  1989      155.9'16      88-35667

Manufactured in the United States of America
Previous volumes in this series available from Grune & Stratton, Inc.
(Harcourt Brace Jovanovich, Publishers)

# Contents

| | |
|---|---|
| List of Illustrations | viii |
| Preface | ix |
| List of Contributors | xv |

**Part I**   **Background and Family Issues**

1  The Body and Personal Integration in Health and Illness   5

   *Leonard S. Zegans*

2  Preparing for Uncertainty: Family Reactions to a Seriously Impaired Child   29

   *Albert J. Solnit*

**Part II**   **Assessment**

3  Psychosocial Evaluation of Physically Disabled Persons   43

   *Mitchell Rosenthal*

4  The Problem for Neuropsychologists: Differentiating Physical, Emotional, and Cognitive Abilities   58

   *Ralph J. Kiernan*

## Part III    Treatment/Rehabilitation

*Childhood Onset, Nonprogressive Disorders: Cerebral Palsy, Spina Bifida*

5  The Influence of Psychosocial Factors on Personality Development and Emotional Health in Children with Cerebral Palsy and Spina Bifida — 87

*Gabriella E. Molnar*

*Childhood Onset, Progressive Disorders: Muscular Dystrophy*

6  Psychosocial Issues in the Treatment of Children with Muscular Dystrophy — 108

*Gloria D. Eng*

*Adult Onset, Nonprogressive Disorders: Spinal Cord Injury*

7  Psychosocial Adjustment to Spinal Cord Injury — 117

*Roberta B. Trieschmann*

*Adult Onset, Progressive Disorders: Multiple Sclerosis*

8  Psychosocial Implications and Treatment of Persons with Multiple Sclerosis — 137

*Sheldon Berrol*

*Facial Disfigurement*

9  Psychological Problems Associated with Facial Disfigurement — 147

*Norman R. Bernstein*

# CONTENTS

| Part IV | **Social Psychological Issues** | |
|---|---|---|
| 10 | Psychosocial Aspects of Assistive Devices for Disabled Persons | 170 |
| | *Jean G. Kohn* | |
| 11 | Sexual Concerns of the Physically Disabled | 183 |
| | *Susan E. Knight* | |
| 12 | The Forensic Psychiatrist's Role in Evaluating Physical Disability in the Workers' Compensation System | 200 |
| | *Carroll M. Brodsky* | |
| 13 | The Poverty of Justice: Ethics and the Physically Disabled | 222 |
| | *Robert M. Nelson* | |
| 14 | A History of the Independent Living Movement: A Founder's Perspective | 231 |
| | *Edward V. Roberts* | |
| 15 | Discrimination and Disability | 245 |
| | *Albert R. Jonsen* | |

Index     253

# Illustrations

**Figures**

| | | |
|---|---|---|
| Figure 3.1 | $B = f(P \times O \times E)$. Behavior (B) as a function of the Person (P), the Organism (O), and the Environment (E) | 46 |
| Figure 4.1 | Schematic Plot of Attention vs. Verbal Learning for 5 Memory Tests | 72 |
| Figure 10.1 | Interaction of the Combination of Person and Device with the Environment | 171 |

**Tables**

| | | |
|---|---|---|
| Table 6.1 | Clinical Course of Duchenne Muscular Dystrophy | 110 |
| Table 6.2 | Therapeutic Management of Children with Duchenne Muscular Dystrophy—Medical, Social, and Psychological Aspects | 112 |
| Table 10.1 | Developmental Aspects of Normal Verbal and Written Communication, and Alternatives for the Nonverbal Child | 174 |
| Table 10.2 | Chronology of Mobility with Reference to Ambulation in Normal and Motorically Handicapped Children | 175 |

# Preface

According to the National Center for Health Statistics (NCHS, 1986), 32.6 million Americans, or about 14.0% of the noninstitutionalized population, are limited in some way by chronic health conditions. More than 9.2 million citizens, or 3.9% of the population, suffer severe disabilities—chronic health conditions that prevent performance of a major activity such as working, attending school, or housekeeping. The number and percentage of Americans suffering physical disabilities has increased dramatically since the turn of the century. In the past 20 years alone, as the population has increased about 22%, the prevalence of severe disability has increased 83%, rising from 213 per 10,000 in 1966 to 390 per 10,000 in 1985. This trend is partly due to medical advances, which have saved the lives of many persons at birth or later in life who would otherwise have surely perished. All too often, however, while death is avoided, disability is not. In addition, dietary improvements, public health programs, and other factors have extended life expectancy at birth from an average of 47 years in 1900 to 74 years in 1980. Disabling conditions, as might be expected, are more prominent among older Americans. Of those over 65, 40% must limit activities because of disability, and 10% are unable to carry on a major activity.

In the past 20 years, the increasing prevalence of disabling conditions has been accompanied by major social and governmental policy changes regarding citizens with disabilities. Increased public awareness, attitudinal shifts, public access, and legislative progress have been generated by disabled persons themselves through political action, education, advocacy, and example (e.g., the independent living movement). At both national and local levels, many laws have been passed to enhance the quality of life of persons with disabilities and to promote their ability to live independently and work in competitive employment. Of the many laws Congress has passed (e.g., PL 90-480, the Architectural Barriers Act of 1968; PL 91-453, the Urban Mass Transit Act of 1970; PL 93-67, the Federal Highway Act of 1973; Pl 94-142, the Education for All Handicapped Children Act of 1975; and PL 95-602, the Rehabilitation Comprehensive Services Act, among others), the jewel in this legislative crown continues to be Public Law 93-112, the Rehabilitation Act of 1973, which bars discrimination on the basis of handicap in any program or activity directly or indirectly benefiting from federal funding. These legislative initiatives cut across conventional liberal and conservative philosophical lines. They are liberal in the sense that they extend civil rights protection to a minority group—citizens who are disabled—and impose some demands upon employers and governmental agencies to remove barriers to full participation in society by persons with disabilities. They are conservative in their emphasis on self-reliance and the right and responsibility of citizens with disabilities to shape their own destinies.

As the number of persons with disabilities has increased, the number of professionals and new techniques, the variety of assistive devices, and the amount of information to help disabled persons deal with physical needs has increased enormously. In addition, the significant and increasing expenditures for disability-related programs have given rise to questions of public policy, calls for cost-benefit analyses, and ethical considerations.

Unfortunately, psychosocial aspects of disability remain poorly addressed (McNett, 1980). To better address these issues, the field must perform a kind of pincers maneuver, building up research and clinical knowledge to enable more effective psychosocial interventions with disabled individuals and their families, as well as developing effective strategies to change public attitudes toward persons with disabilities. In one sense, the second task is the harder one: it is easier to enable an individual to come to terms with his or her disability given an environment that is not sympathetic, than it is to change the attitudes of an entire society. Nevertheless, it is essential to work on both fronts since the environment plays a major role in the success of treatment and rehabilitation. Despite increased public sensitivity, consumer advocacy, and legal measures, the main barrier to full social participation by persons with disabilities remains the view that "temporarily able-bodied" people have of disabilities and disabled persons. As DeJong and Lifechez (1983, 49) point out: "The ultimate and most pervasive of environmental barriers are the attitudinal ones, particularly the view that disabled people are helpless, pathetic victims, deserving of charitable intervention. There is now more than enough experience to indicate that disabled people can, with appropriate environmental supports, lead full and independent lives. Without the removal of attitudinal barriers the disability legislation of the past decade will not realize its full promise."

The difficulty such an environment poses for a person with a severe disability—even with a healthy, cohesive, supportive family—is vividly evoked in the autobiography of a gifted young Irish poet, Christopher Nolan, who suffers from cerebral palsy. Though near-death from asphyxiation at birth robbed him of movement and speech, it did not steal his intelligence or his passions—in fact, it may have heightened them. Nolan eventually learned to communicate by punching a keyboard, one letter at a time, with a rod attached to his forehead, as his mother held his chin in her hands to steady it. In this way, painfully slow and awkward though it was, he could at last communicate with the world! One result of the flood of words released by this event was a critically acclaimed book of poetry, *Dam-Burst of Dreams,* published at the age of 15. Another was the growth of his own identity, feelings of self-worth, and self-confidence, which led him—with the support of his family—to actively, and ultimately successfully, challenge the unsympathetic social environment.

And so the battle was staged between a crippled, sane boy and a hostile, sane, secretly savage though sometimes merciful world. Can I climb man-made mountains, questioned Joseph Meehan [Nolan's name in the book]. Can I climb socially constructed barriers? Can I ask my family to back me when I know something more than they, I now know the heinous skepticism so kneaded down constantly in my busy sad world. What can a crippled, speechless boy do, asked Joseph, my handicap curtails my collective conscience, obliterates my voice, beckons ridicule of my smile and damns my chances of being accepted as normal. (Nolan, 1987, 4)

Clearly, in addition to the physical disabling condition per se, emotional, social, and vocational concerns must be addressed before, during, and after treatment and rehabilitation by patients, their families, and the professionals who serve them. Sadly, knowledge and training in this area have not kept pace with trends in the physical and biological arenas, though recent advances and the establishment of fields such as health psychology and behavioral medicine seek to address these concurrent issues (Stone et al., 1979; Weiss, 1983). The need for an integrative approach to studying disability, however, remains largely unmet. It was to assist in remedying this situation that the present volume came into being.

This fourth volume in the Mind and Medicine series is a companion to the previously published *Psychosocial Interventions with Sensorially Disabled Persons* (Heller et al., 1987) and addresses salient issues in the psychosocial assessment, treatment, and rehabilitation of persons with physical disabilities. It offers theoretical, research-based, and clinical information useful for professionals in rehabilitation, nursing, mental health, and medicine, who in partnership with disabled persons and their families, compose the treatment/rehabilitation team. Each chapter is written by an expert in the field who is usually a pioneer in his or her specialty as well. Each editor has had personal and professional experiences that have profoundly affected his view of the subject. Louis Flohr has experienced the rage and despair at the birth of a severely impaired child and the profound sense of grief and relief when she mercifully died. The deep sorrow that remains has both informed and motivated him in his work on this volume. Leonard Zegans spent time as a medical student at the Rusk Institute in New York, thereby gaining a firsthand impression of the connections between mind and body that are salient for each person with a physical disability. And Bruce Heller served as director of training of the Center on Deafness at the University of California, San Francisco for seven years and is currently a consulting psychologist to the Oro-Facial Institute, Pacific Medical Center. In these capacities and in his private practice, he has been profoundly impressed by both the many frustrations and impediments occasioned by disabling conditions as well as the strength of character and ingenuity with which many individuals overcome them.

The first section of the book, "Background and Family Issues," includes an overview of the "mind-body" problem as it relates to disability and a discussion of parental reactions to the uncertainty that accompanies the birth of a seriously impaired child. Chapters in the second section, "Assessment," address issues in the psychological, social, and neuropsychological evaluation of persons who are disabled.

Given the impossibility of discussing every disability, the third section, "Treatment/Rehabilitation," employs four major syndromes as paradigmatic of issues that are encountered in four categories of disabilities defined along two dimensions: age at onset and course (i.e., whether the disability first occurred in childhood or adulthood and whether the disabling condition is progressive or nonprogressive). In addition, one chapter addresses the important, compelling, but often neglected area of facial disfigurement. The final section, "Social Psychological Issues," offers perspectives on psychosocial aspects of assistive devices, sexual issues related to disability, forensic psychiatric issues, independent living (from the founder of the independent living movement), and ethical aspects of disability.

This book arose from specific assumptions and incorporates several implicit perspectives. It is intended to be practical but also to integrate theory, research findings, and clinical wisdom. It focuses on persons rather than on disabilities or diseases per se, and it holds that the most useful locus of intervention is not the disability but the person. It embraces an interpersonal, developmental, and holistic perspective, finding in the relationship between the health professional and the person with a disability a most powerful curative and rehabilitative agent. This book, through its contributors, espouses a multidisciplinary approach in the belief that the most effective intervention is potentiated by a team of professionals from various specialties working together and in partnership with patients and their families. The authors seek to debunk myths, including the results of "psychologizing," which often seem to "blame the victim" rather than assist the disabled patient. Yet they also espouse addressing psychosocial issues as crucial in the treatment/rehabilitation process. This book eschews purity of theory or orientation in favor of interventions that work. We hope it will stimulate further thought.

B.W.H.<br>
L.M.F.<br>
L.S.Z.

## REFERENCES

DeJong, G. & Lifchez, R. (1983). Physical disability and public policy. *Scientific American, 48,* 40–49.

Heller, B. W., Flohr, L. M. & Zegans, L. S. (Eds.). (1987). *Psychosocial interventions with sensorially disabled persons.* New York: Grune & Stratton.

McNett, I. (1980). Mental health services for handicapped fall between agencies. *American Psychological Association Monitor, 11,* 4, 31.

National Center for Health Statistics. (1986). Current estimates from the National Health Interview Survey, United States, 1985. By A. J. Moss & V. L. Parsons. *Vital and Health Statistics,* ser. 10, no. 160. DHHS pub. no. (PHS) 86–1588.

Nolan, C. (1987). *Under the eye of the clock.* New York: St. Martin's.

Stone, G. C., Cohen, F. & Adler, N. E. (Eds.). (1979). *Health Psychology: A handbook.* San Francisco: Jossey-Bass.

Weiss, S. (1983). Health and illness: The behavioral medicine perspective. In L. Temoshok, C. Van Dyke & L. S. Zegans (Eds.), *Emotions in health and illness.* New York: Grune & Stratton.

# Contributors

**Norman R. Bernstein, M.D.**

Professor, Department of Psychiatry, School of Medicine, Harvard University, Cambridge, Massachusetts

**Sheldon Berrol, M.D.**

Associate Clinical Professor, Physical Medicine and Rehabilitation, School of Medicine, University of California, San Francisco, San Francisco, California

**Carroll M. Brodsky, M.D., Ph.D.**

Professor, Department of Psychiatry, School of Medicine, University of California, San Francisco, San Francisco, California

**Gloria D. Eng, M.D.**

Professor of Child Health and Development and Medicine, George Washington University Medical Center; and Director of Physical Medicine and Rehabilitation, Children's Hospital National Medical Center, Washington, D.C.

**Louis M. Flohr, M.D.**

Assistant Clinical Professor and Director, Disabilities Training Program in Child Psychiatry, Department of Psychiatry, School of Medicine, University of California, San Francisco, San Francisco, California

**Bruce W. Heller, Ph.D.**

Assistant Clinical Professor, Department of Psychiatry, School of Medicine, University of California, San Francisco, San Francisco, California

**Albert R. Jonsen, Ph.D.**

Professor and Chair, Department of Medical History and Ethics, School of Medicine, University of Washington, Seattle, Washington

**Ralph J. Kiernan, Ph.D.**

Assistant Clinical Professor, Department of Psychiatry, School of Medicine, University of California, San Francisco; and Department of Psychiatry, School of Medicine, Stanford University, Stanford, California

**Susan E. Knight, M.S.W.**

Lecturer, Department of Psychology, John F. Kennedy University, Orinda, California; and past Director, Sex and Disability Unit, Human Sexuality Program, Department of Psychiatry, School of Medicine, University of California, San Francisco, San Francisco, California

**Jean G. Kohn, M.D., M.P.H.**

Adjunct Lecturer, Maternal and Child Health Program, School of Public Health, University of California, Berkeley; and Medical Advisor, Rehabilitation Engineering Center, Children's Hospital, Stanford Univeristy, Palo Alto, California

**Gabriella E. Molnar, M.D.**

Clinical Professor, Department of Rehabilitation Medicine, School of Medicine, University of California, Davis; and Director, Department of Pediatric Rehabilitation, Children's Hospital Medical Center of Northern California, Oakland, California

**Robert M. Nelson, M.D.**

Assistant Professor, Department of Pediatrics, Division of Critical Care Medicine, School of Medicine, University of California, San Francisco, San Francisco, California

**Edward V. Roberts**

President, World Institute on Disability, Berkeley, California; past Director, Department of Rehabilitation, State of California; and MacArthur Award Recipient

**Mitchell Rosenthal, Ph.D.**

Assistant Professor, Department of Physical Medicine and Rehabilitation, Rush Medical College, Chicago, Illinois; and Director, Psychological Medicine, Marianjoy Rehabilitation Center, Wheaton, Illinois

# CONTRIBUTORS

**Albert J. Solnit, M.D.**

Sterling Professor of Pediatrics and Psychiatry, School of Medicine and Child Study Center, Yale University, New Haven, Connecticut

**Roberta B. Trieschmann, Ph.D.**

Consulting Psychologist, Private Practice, Scottsdale, Arizona

**Leonard S. Zegans, M.D.**

Professor and Director of Education, Department of Psychiatry, School of Medicine, University of California, San Francisco, San Francisco, California

# PART I

## Background and Family Issues

The chapters in part one offer background information and provide a context for the rest of the volume by reviewing how the body has been perceived by medicine, rehabilitation, and philosophy and discussing the issues that may arise for family members and professionals at the birth of a seriously impaired child.

Leonard Zegans, in his phenomenological overview of the mind-body problem as it relates to physical disability, characterizes the body as "an inseparable and intimate attribute of . . . self." He describes the importance of the body and its relationship to the self as seen by physicians, psychotherapists, philosophers, and biologists. Modern medicine, Zegans states, has had much success pursuing a materialistic, positivistic program of research and treatment. Paradoxically, because of these successes, medicine holds an overly mechanistic view of the body, which will impede even greater success in the future. It leads practitioners to focus on the disease or disability and the organs affected rather than to treat the person beyond the target condition. Failure to recognize the unity of body and mind and, therefore, to treat the whole person may hinder recovery and rehabilitation.

Zegans argues against the "medical model" in which the body is regarded as a "passive citadel under attack by destructive forces." Instead, he offers a phenomenological perspective, which does not discard materialism, positivism, and the scientific method but integrates the view of the body as a thing with the view of the body as an integral attribute of the self. In this sense, the self is perceived as a larger entity, superordinate to the body and to physical disability. Such a synthesis, he suggests, enables professionals and patients to deal with disabilities more fruitfully and offers the person with a disability a potentially richer life. "Disability," he concludes, "need not be a limitation for a healthy and expansive self."

Family issues figure prominently in the treatment and rehabilitation of persons with disabilities. Albert Solnit's chapter continues his classic discussion of the grief parents experience when, rather than the healthy child they expected, a disabled child is born. He focuses on the pervasive uncertainty generated by such a birth in both parents and sibs as well as professionals. In one sense, as Solnit points out, it is impossible to prepare for uncertainty since, by definition, uncertainty implies unpredictability. However, families with healthy, resilient relationships cope with uncertainty better than do those without such interpersonal resources and skills. Even families with robust coping mechanisms, however, may find themselves initially at a loss and will need time to accept the reality of such a birth, to work through their own anguish, and to marshal coping strategies.

Whether coping skills are good or poor, however, Solnit cautions professionals not to impose their own views and services in a preemptory way as a result of their own anxiety, uncertainty, and feelings of powerlessness. Instead, they should seek to strengthen the family's coping abilities as it works through its grief and begins to make decisions about the future. Differences in the tempo and rhythm of intervention between family and professionals should not be allowed to interfere with the coordination of care to the detriment of the child. Solnit discusses the need for professionals and parents to develop good working relationships since both are essential collaborators in the therapeutic and rehabilitative effort. He encourages parents and family members to assert their opinions, to ask questions, and to feel free to disagree with professionals, and he encourages professionals to be responsive to such queries and the anxiety fueling them. He suggests that professionals take an attitude of cautious, reserved optimism. Such an attitude is least harmful in the sense that it acknowledges professionals' limited prognostic power and, in addition, adds a feather's weight of optimism—a communication of hope for and commitment to life—that will most likely be beneficial, whatever the outcome.

# 1

# The Body and Personal Integration in Health and Illness

*Leonard S. Zegans*

More than a few years ago as a medical student, I worked with disabled patients at the Institute for Physical Medicine and Rehabilitation in New York. A stroke patient, who had been vigorous and successful in business, helped me to think about the body in illness differently from how I had been taught in physiology and pathology courses. His stroke made him reexperience his sense of his body, which for most of his life he had taken for granted. It had been most intimately his when he didn't think about it at all. Now, he reflected, he thought not only about the parts of his body that he could not move but about all of the parts. What once seemed a unity now appeared like a collection of fragmented components. There was, he said, "a separation between me and my body, which was new and frightening." He felt as if he no longer inhabited his body, and this, even more than the specific handicaps of movement, was the greatest loss of all. What he was implying was that his body had been for him more than a set of tools or instruments for performing certain functions: it was an inseparable and intimate attribute of his self. Disability cost him the unacknowledged harmony of his body.

The issues of disability dealt with in this volume bring into vivid focus the manner in which the body is regarded by patients, clinicians, and science. For all of us, our bodies are in some ways the most intimately known aspect of our selves, yet, also mysterious and conflictual. At times our bodies perceive and smoothly perform complex actions without our being aware how we are enacting such intricate behaviors. During these moments there appears to exist a meshing of motivation and action, an unreflective *doing,* when our will, our behavior, and our environment are seamlessly combined. On other occasions we experience our bodies as distant, dead, a thing that is unresponsive to our wishes, an unfamiliar object that has nothing to do with our basic sense of self. We ignore our body and at times disparage it, yet when it does not function as we expect, it evokes in us profound anxiety and dread.

Throughout history no topic has seemed as inexhaustible as the variety of ways we exalt, condemn, observe, and embellish our bodies. Our theories and models about our bodies often picture them as either intricate machines or sanctified reflections of divinity. Helping professionals constantly seek to understand the relationship between patients and their bodies. Despite repeated injunctions to "treat the whole person," many professionals become experts of a certain organ system, disease, or syndrome, often forgetting the essential unity of body and self. There lingers in medicine and the behavioral sciences the tendency to separate body and mind, to fragment this unity, while dealing like technicians with the human corpus (Zegans & Victor, 1985).

When physicians confront a disease, their attention invariably focuses on its pathogenesis and course, the modes of intervention, and probable residual effects. There is an emphasis on reducing structural organ damage by using the proper therapeutic and recuperative regimes. Despite recent awareness of host resistance factors, modern medicine tends to view the body in illness as a passive citadel under attack by destructive forces. The body and its organs are conceived of as targets of battle to be guarded from the assaults of bacteria, viruses, cancer cells, and autoimmune antibodies and then repaired. Proper strategy is established by bringing in outside forces (e.g., antibiotics, laser beams, artificial organs, cobalt rays) until the body can rally its own defenses. Increasingly, physicians comprehend the body from the framework of complex molecular processes including ion exchange, receptor sensitivity, synaptic uptake, and enzyme kinetics. This knowledge, while enhancing the range and effectiveness of treatment, has deepened our understanding of how the body operates. Yet we tend to forget that for its owner the body is, in a larger sense, a vessel of meaning, memory, and intention.

As we grow, we acquire a sense of our body as it relates to the environment and to the self. Images and feelings concerning our body are built up through a variety of personal and social experiences. Attitudes about the body and its parts are conveyed to us by parents, relatives, friends, teachers, and religious figures. In time our body can become a sign of our status, power, beauty, desirability, or competence. Some of us learn to respect the body and care for it, while others regard it with distrust, revulsion, or despair.

By the time we become the victim of an injury, illness, or infirmity caused by aging, certain attitudes toward our body have already been established within us. We build up schemata about posture and perception; parts of our body we may value, while others are the pivot of unconscious conflict. There are aspects of our body to which we may never pay attention (such as breathing or heartbeat), while a minor skin blemish can become the center of anxious concern. Our body also develops as a medium of expression of feelings and motivations. At times our facial expressions, posture, and gestures are woven into a rich tapestry of communication, yet at other moments our body is an inert mechanism. Especially when working with disabled patients, it is critical that caretakers understand how patients perceive their bodies. Caretakers must appreciate the meanings and apprehensions patients attach to both injured and healthy parts. As they more fully comprehend their patients' sense of self and attitude toward the body, its functions, and its boundaries, practitioners become more empathetic and effective agents of recovery.

In this chapter I will explore different ways in which the body has been regarded by physicians, psychotherapists, philosophers, and biologists. I will argue that though modern medicine has made outstanding advances in understanding the underlying mechanisms of human physiology—advances which have revolutionized health care—it continues to hold a mechanistic or materialistic view of the human body. It relies on an incomplete portrait of the human body and leaves unexplored rich aspects of human embodiment. During this century, however, a number of philosophers and practitioners have come forward with different ideas about the body in health and illness. These phenomenological or psychological models do not reject the rich insights of the mainstream biomedical approach, but aim at a new vision of the body, a vision that simultaneously embraces the body's thinglike objectivity and its integral participation in self. Such a model attempts to overcome the dualism that views the human as a machinelike body separate from an extentionless mind. An integrative model is necessary to help patients deal with their disabilities and reach a rich enactment of their selves.

I will expand on these themes in the sections to follow. First I will discuss various ways in which conflicts, fantasies, and fears about the body are observed during the course of psychotherapy. Then I will explore what is meant by the self and how it is important in thinking about disability. In the next section, on embodiment—the integration of one's physical and psychological functions—I will argue that the unfolding self can provide guidance and continuity for the ailing, healthy, or developing body. The patient's sense of his or her power and competence will then be examined along with plasticity, the capacity to reorganize structure and functions in the face of damage or loss.

The chapter concludes by positing that mind and matter are always present together in a taut relationship. When mind neglects body, it loses its balance and measure, because it is through the body that we make contact with the self. Disability need not be a limitation for a healthy and expansive self.

The various papers that are presented in this volume and its companion show us how we can bridge the conceptual and practical gaps between paradoxically different models for treatment, when we work with patients who have either physical or sensory disabilities.

## PSYCHODYNAMIC ATTITUDES TOWARD THE BODY

As a psychiatrist and a physician I have been enabled to learn a number of things about how patients think about and discuss their bodies. Many of the patients with whom I work are not physically ill or disabled, yet all of them have had critical concerns about their bodies, which have influenced their reactions to both physical and mental disturbances. Although themes that focus on the body often occur during psychotherapy sessions, they are frequently avoided or dismissed by therapists. Clinicians with psychodynamic training often prefer to discuss conflicts, object relations, and emotions with patients as though these topics occur in an abstract sphere, detached from the realities (that Freud so often emphasized) of corporeality. In any somatic or mental illness, the caretaker must confront the role the body plays in the affective, cognitive, and motivative life of each patient. Bodily themes that emerge during psychotherapy represent universal concerns, fantasies, and wishes that can be observed in a variety of other clinical settings. With ill or disabled patients, the physical therapist, genetic counselor, special education teacher, psychotherapist, and primary physician must be sensitive to patients' ideas about the body that affect their self-esteem and motivation. Often unspoken yet conscious worries about the body can interfere directly with physiological processes of healing and with the patient's cooperation during treatment.

I have divided those themes that emerge frequently in my work with patients into four broad categories:

# THE BODY IN HEALTH AND ILLNESS 9

1. The body as a focus of psychological experience including desires, fears, and conflicts;
2. The body as a vehicle (or medium) for social interactions;
3. The body as a symbol of the person's value, desirability, and competence;
4. The body as a symbol of the person's psychophysiological integration of the self.

Awareness of these themes is important for those working clinically with disabled persons, since they affect patient response in therapeutic and rehabilitative efforts. Each of these categories will be illustrated below.

## The body as a focus of psychological experience including desires, fears, and conflicts

*Conversion reactions*

Psychiatrists continue to see patients for whom various functional disabilities such as paralysis or sensory defect arise as the result of a reaction symbolically expressing through an organ system some deep intrapsychic conflict. A body part is used to express some sexual or aggressive impulse that is blocked from external expression. Conversion disorders show us that any part of the body may come to psychologically represent some other part, function, or impulse.

*Psychosomatic ailments*

While in conversion reactions, the body does not function properly despite the absence of any structural damage, stress and tension may cause or contribute to a wide range of conditions in which pathological changes in organs do occur. Whether such illnesses represent specific expressions of intrapsychic conflicts or more general failures of coping and adaptation under conditions of stress is currently debated. What is clear is that emotions can contribute to physiological conditions that make the body more vulnerable to injury or disease and less resistant to the effects of established illness.

*Idiosyncratic ideas held by the patient about body functioning*

Therapists often have patients who tell them thoughts about their bodies that seem unrealistic or bizarre. They may imagine that parts of their bodies are under another person's control, that certain organs are missing or duplicated, or they may attribute special sensory or motor powers to themselves.

Often, wishes about other persons are transferred to their own bodies, where different organs or tissues enact a fantasized, interpersonal drama. In the course of treatment, the patient may express ideas about an organ's location, appearance, and function that represent psychologically important fantasies that can be traced back to childhood. Such ideas persist into adulthood and affect the person's feelings about bodily functions, potential dangers, and vulnerability to illness. The private view of one's body, how it looks and operates, tells a great deal about one's wishes, fears, and sense of integration.

*Body "armature"*

Many therapists (starting with Wilhelm Reich, 1949) have been concerned with the way in which the muscles of the body at rest or in motion reflect past tensions and conflicts. Postural tonus can be a sign that some part of the body has been actually or symbolically traumatized in the past. Certain muscle systems (shoulders, pelvis, chest, diaphragm, legs, face) may appear constricted or frozen. The body may communicate in this static mode a burden of pent-up anger, frustration, defeat, and fear. At times, reflecting past conflicts, the patient may be unable to orient his or her body correctly in space and displays uncertainty and clumsiness. Focusing therapeutic attention on muscular tension can release frozen memories and affects associated with conflictual past events.

*Fears centering on bodily destruction, mutilation, or malfunctioning*

Frequently in the dreams, free associations, and fantasy productions of patients, therapists learn of vivid fears of bodily danger. These ideas often trace back to early childhood when patients felt vulnerable to the assaults, real or imagined, of others. Any illness or injury can vividly bring back primitive fears of bodily harm. Concern with bodily integrity and competence of function is close to most of us. Yet patients often do not express their anxieties when being told of an illness or while undergoing some medical procedure. Frequently patients do not share thoughts relating to loss of health because of the intense anxiety this arouses.

*Desire for self-mutilation*

This concern is often the mirror image of the fear of bodily damage. The patient may hate a body part and wish to do away with it. The part may represent some conflict or be the symbol of a hated person; it may represent the patient's general feelings of futility and inadequacy, and damage to it may actually be a defense against self-destruction. In illness or injury the patient might view the part as something foreign. A damaged organ may be personified as something external to the core of identity. The neurologist Critchley (1955) talks about stroke patients giving names to impaired limbs: "old useless," "the delinquent," "gammy," or "the nuisance" may be used. Assisting a patient in integrating a damaged or diseased body part back into the total self-schema is a difficult task for any clinician. Yet the patient's wish to renounce or destroy the despised member may further slow healing and recovery.

*Libidinization of body parts*

It is not unusual to have a patient take a sudden interest in a part of the body that is given new and inappropriate meaning. Sexual interest can be shifted away from the eroticized "other" and from the genitals onto a different body part. This, of course, may happen in a conflictual and unconscious manner in conversion reactions. In this instance, the body is not being eroticized in the service of interpersonal sexuality, but as a substitution for it, a withdrawal. The patient may thereby be expressing very early states of bodily eroticization existing prior to the genital consolidation of puberty. On the other hand, this plasticity of the central nervous system (which is the ultimate organ of sexuality) serves a highly adaptive function in disabilities (e.g. spinal cord injury) through which the patient loses genital sexuality and wishes to learn alternative means to reaching orgasm.

*Projection of fear about bodily impairment*
*on other persons or objects*

The patient who is in conflict about some bodily function, structure, or impairment may deal with the ambivalence by projecting concern onto others. A patient who is losing his or her looks may become preoccupied with the wrinkles, grayness, and loss of skin tone seen in acquaintances. Therapists who suddenly hear about the deformities, inadequacies, and loss of bodily control of others must consider the chance that patients may be communicating their own bodily worries, which may lie outside their conscious reach.

*Bodily transformation as reflected in*
*dreams, daydreams, and fantasies*

If, as Freud (1964) stated, the ego is rooted in the body, so too are our unconscious impulses, as well as our moral reactions against them. The size, shape, function, and transformation of body parts are like a sign language for our imagery. Therapists can gain much understanding of their patients through a "reading" of body imagery, which comes from outside the discursive language of the patient. The wishes, fears, and conflicts of the patient often present to the therapist through the syntax of body images. Any alteration in bodily status, whether through puberty, pregnancy, illness, surgery, or aging, will likely release certain images of the body in some symbolic production.

## The body as a vehicle (or medium) for social interactions

### Issues centering on the permeability or rigidity of the body boundary

Concern about personal boundaries often reflects patients' conflictual integration of their body image with their sense of self. Patients experiencing individuation and separation problems may have difficulties effectively establishing the borders between themselves and others. They may fear merging with others or being penetrated or engulfed. Sometimes these patients become excessively resistant to any attempt by others to come close or touch them because of fears concerning the lack of demarcation between self and surroundings. Body-boundary issues for patients who have recently become ill or disabled are very important both in physical and psychotherapeutic work.

### Problems centering on the use of the body as an expressive vehicle in interpersonal communication

We use our body to express our feelings and intentions to others, and we also perceive the expressive repertoire of others. This is vital to the establishment of healthy object relations, a sense of reality, and feelings of personal adequacy. Obviously in cases of sensory disability or when there is facial or bodily disfigurement, such anatomical defects can interfere with the expressive use of the body. Yet patients may also have difficulty using the body expressively because of unconscious conflicts or because of inadequate early social learning. Problems in the expressive use of the body isolate the individual and reduce vital affective communication with others. The clinician may notice a dissociation between words expressing feelings and the nonverbal cues accompanying them. Certain emotions may never be expressed even though the circumstances call for or permit them. Some emotions are expressed in either attenuated or exaggerated fashion, and the person may seem embarrassed and apologetic for exhibiting feelings openly. Patients may appear uncomfortable and graceless in their movements. Their bodies appear disarticulated in action and gesture, without fluidity or rhythmic balance. The gestures show gaps, hesitations, and erratic flow, often appearing more related to internal states of anxiety and tension than to a clear, intentional focus on interpersonal communication.

*Concerns about the interest of others in the patient's body*

Almost as emotion ladened as the patients' concern about their bodies is their elaboration of ideas about how others view and wish to use their bodies. Considerable effort in children and adolescents goes into differentiating their bodies from those of their parents and learning what bodily activities can be safely and gratifyingly shared with others. Children wish to have others admire and desire their bodies but also fear manipulation and control. Significant struggles in adolescence and childhood revolve around the parent's wish to prescribe the use of the offspring's body. Battles around movement, eating, elimination, and eroticism focus on the child's wish to determine how, where, and when bodily functions can be exercised. Superego formation begins with the child's gradual assumption of parental imperatives about regulating body use. Children desire the free use of their bodies while also wishing to please their parents by internalizing parental attitudes toward the body. They seek their parents' admiration and approval of their bodies but are both excited and disturbed by parents' erotic, aggressive, or appropriative interest in these bodies. Somatic disturbance amplifies these themes of freedom, control, interest, appropriation, and misuse of one's body by another.

## The body as a symbol of the person's value, desirability, and competence

*Interest in the body parts of others in comparison with one's own*

Throughout life one's sense of self-worth is in part entwined with one's body. There is a constant comparison of shape, size, texture, and color of one's own body with that of others. A feeling of adequacy, shame, or worth often arises from a sense of the comparative value of one's body. Even as the development of the mind and cognition come to play an important part in our lives, we constantly appraise our corporal worth in comparison with others'. In injury, a loss of function or disfigurement revives and accentuates these concerns whether consciously or not.

*Focus on moral feelings about the body*

Patients' feelings of moral disapproval of body parts or functions can be seen at any point during the course of treatment. Disapprobation may be centered on body parts and processes that are regarded as wrongfully sexual or aggressive. Rekindling of such concerns is often seen during an illness resulting in disability, where affliction may be regarded (consciously or unconsciously) as punishment for the abuse of the body. Patients may be able to experience their bodies intensely, but such an experience is occasionally fraught with feelings of being sinful or out of control. They are unable to feel comfortable with bodily sensations yet are often fascinated by them. Guilt and shame can ultimately lead to a sense of self-denial and isolation from the body.

*Frequent changes in body appearance*

Many parts of the body can be adorned and altered. Patients sometimes show their discomfort with body image by frequent changes in hairstyle, dress, body ornamentation, weight, and height. Willingness to experiment with one's looks can be a positive sign of a capacity to attend to the body. But rapid and, at times, bizarre shifts in the presentation of the body can also be a clue to some difficulty in accepting one's appearance to oneself and to others. The body is the bridge between ourselves and the outside world. The tension, guilt, and discomfort we experience toward our bodies are invariably communicated to others. An alienation between mental and corporeal aspects of self can be "contagious," interfering not only with our capacity for empathy with others, but also with our ability to engage their interest and support.

## The body as a symbol of the person's psychophysiological integration of the self

*Depersonalization*

There is a wide range of phenomena, often communicated by patients in therapy, which indicate a sense of separation between their experience of self and their bodies. Feeling disembodied is a common experience in schizophrenia but can also be observed with borderline patients, neurotics, and normal individuals undergoing a strong emotional upheaval. Depersonalized states can also be observed with patients abusing alcohol or other substances.

*Feelings of the patient about personal body image or schema*

The concept of the body image or schema (to be discussed in more detail below) refers to the patient's sense of the total body, how each part articulates with the others, and the relationship of the body to the environment. We are usually not consciously aware of our body schemata as we make different postural adjustments to our surroundings. However, during times of illness or injury, we become exquisitely sensitive to changes in our image of the body and how it relates to our world of action. Concerns about body schema are also seen in patients undergoing important life transitions, e.g., adolescence, pregnancy, and aging.

*Issues centering on misrepresentation of body size and shape*

Damage to the parietal lobes often can produce disturbances of morphology and bodily function. Yet patients with no brain damage may also over- or underestimate the size of body members and their shapes in response to a variety of psychological influences. Patients may exaggerate the size of their sexual organs, overall weight, or height as a response to some feeling of inadequacy or internal conflict about their bodies. Problems in ego maturity may be reflected in both this category and the issue of body boundary. As Freud (1964) pointed out, "The ego is first and foremost a bodily ego; it is not merely a surface entity, but is itself a projection of a surface. The person's own body, and above all, its surface, is a place from which both external and internal perceptions may spring." Sudden changes in a patient's perception of his or her body schema, body boundaries, or the shape and size of body components are always of extreme clinical importance, indicating either some organic problem or an alteration in equilibrium between ego and unconscious impulses.

*Hyper- or Hypoawareness of body parts or functions*
*(e.g. concentration on breathing, heartbeat, and bowel movements)*

During times of stress or heightened emotions, patients may become aware of their bodies in different ways. Motions or processes that normally fall outside the range of consciousness may become obsessively centered in the patients' thoughts. Complex actions become broken up into components, impairing their smooth and flexible performance. Yet, just as it is disruptive to fixate inappropriately on aspects of the body, so too is it maladaptive to be unresponsive to bodily cues. The patient who is unable to "read out" autonomic responses in provocative situations may be unable to make necessary appraisals and responses. Illness and injury alter our usual patterns of body awareness. We may become excessively attuned to insignificant cues, giving them an attention and importance that break the unity and rhythm of our mental and physical processes. However, stress from illness or injury can also lead us to ignore important bodily signs and to isolate ourselves from our corporeal being, thus exposing us to health dangers. Both overattention and negligence of the body are potential dangers that should be explored with the patient.

*Representation of the self in body parts or functions:*
*The body as a symbol of concern about mortality,*
*change, or mutability and of human existence*

How individuals regard themselves as a totality is often reflected in body presentation—its posture, muscle tone, color, liveliness of gesture, facial expressiveness, and sense of warmth or coldness. It is important for the clinician to note whether there is a feeling of unity between the psychic and corporeal aspects of the patient or whether mind and body seem disarticulated and alienated from each other. Sometimes a patient may identify the self with only one part of the body, focusing narcissistic attention on that feature while neglecting the rest. When this part is injured or incapacitated, the psychic wound may be especially hard to bear. Bodily changes brought about by either age, illness, or disability may highlight broad questions about human mortality, the transience of the stages of life, and issues of ultimate meaning. Concerns about disease and bodily decay often initiate a spiritual crisis that brings about a disinterest in the "flesh," ascetic habits, rumination about past sensuous indulgences (real or imaginary), and greater involvement with religious or ethical teachings. Certain patients will report a lack of rapport with their own bodies. They achieve little pleasure from activities that should be sensuously gratifying, and they experience an isolation from significant bodily cues. They sometimes cannot tell when they are hungry, satiated, sleepy, or in pain. Instead of an enlivened corporeal presence, the body is experienced as a mechanical instrument of their will.

The issues raised in the preceding discussion illustrate how the body can be a focus of denial, conflict, fantasy, guilt, and shame for a patient. Clinical work continually illustrates that serious schisms can arise between a person's psychological and somatic processes. The human mind is unique in the way its imagery, rationalization, and capacity for self-deception can become dissociated from the real needs of the body. These psychological processes often occur during development, that is, prior to a major illness or disability. A serious ailment can amplify and exaggerate conflicts about the body and its functions, thereby impairing the ego and threatening the sense of self.

## BODY AND SELF

In any discussion of illness and disability, the importance of restoring self-confidence, self-worth, and self-esteem to the patient arises. Yet, what is this "self" that is to be esteemed? Many philosophers and psychologists regard it as simply a manner of speech, a "ghost in the machine." The Scottish philosopher David Hume declared that the self was no more than the "perception of a succession of bundles of stimuli," a mere grammatical fiction (Hume, 1739). In our time, the Harvard psychologist Gordon Allport (1955) assured us that while "all the psychological functions commonly ascribed to as a self must be admitted as data, what is *unnecessary* and *inadmissible* is the self that is said to perform acts, to solve problems and to steer conduct in a trans-psychological manner."

Another, longer tradition, however, has taken the self as a living, meaningful reality. In ancient Greece, Socrates admonished men to "know thy self," and Pindar advised, "Choose thyself." Shakespeare's words from Hamlet still resonate for us today. "To thine own self be true." These poets and philosophers affirmed that behind the outward manifestations of human behavior, there was a core of being that acted voluntarily and purposefully. Whether they called it "self," "soul," or even "central integrating principle," there appears to be an irreducible human core, which is the ground of human action and values (Zegans, 1982).

I regard the self as a system of personal integration for perceiving and solving the problems of living. It is a system of order, coherence, and openness to change. The self is faced with the task of maintaining stability, coherence, and continuity of form while being capable of transforming, rearranging, and developing psychic structures to permit adaptive response to environmental and maturational challenges. The self selects those drives, objects, defenses, perceptual sensitivities, and modes of communication that best serve our integrative and individuating needs. It endows us with an evolving *meaning structure,* which interprets events, unifies elements in the personality, and provides the basis for action. This is done in a manner that reveals a signature, a style of organization, which we recognize as our self. Thus, individuals have both a sense of what properly fits and reflects the self and what is unassimilated or alienating. Of importance for our current considerations, self is not a separation of the knowing mind as subject from the body as known object. The whole body is to be regarded as the bearer of implicit consciousness. The integration of body and volition plays a strong role in the thinking of both existentialist philosophers and psychodynamic therapists. Does my body belong to me? Am I identified with it? Do I treat my body as an object? Am I enslaved by it? Do I experience my body as separate from myself, so that with its death my essence will live on? Our experience as an embodied self is critical to our capacity of intersubjectivity, the ability of one self to know another, and through the other to better know one's self (Idler, 1979).

The philosopher Merleau-Ponty (1964) places the body at the heart of intersubjectivity. He avers that we have no direct access to the psyche of others (to others' feelings of their own existence), nor can we know their experience of their own bodies. And yet, we do know others through their bodies and ours. People project their experiences of their own bodies onto others. When we witness the bodies and actions of others, we find in them a meaning, because they are the themes of possible activity for our own bodily development.

The self strives toward equilibrium and unity but is capable of change when confronted with new stresses and demands. It constantly seeks new levels of coherent stability, which permit not only environmental adaptation and control but new forms of creative expression and optimal performance. The self uniquely accepts intervals of instability in order to incorporate new information into richer, more effective organismic structures. This parallels events in psychotherapy where older, maladaptive structures are abandoned, creating periods of unease. Ideally the self then moves toward more complex and empowering states of integration and richer modes of personal expression.

Changes in the body brought on by normal maturation and decline or by illness alter the sense of self and create demand for a new integration. An illness affects not only specific tissues or organs but also challenges the coherence and stability of the existent self. The experience of many clinicians who have worked closely with disabled patients is that the self can grow creatively in spite of bodily loss but that it is vulnerable during illness, since the body and its meanings are integral to self-organization.

It is clinically worthwhile assuming that most patients have evolved a personal construct or myth about their bodies in relation to self. The myth may touch on issues concerning the body's appeal, size, value, boundaries, or function. Part of this construct may lie outside of awareness and require special therapeutic approaches to reveal. Much, however, is consciously available to patients and will affect what they think about their illnesses and the approaches to treating it. A perceived loss or malfunction of the body will be interpreted within the bounds of that myth, and that interpretation will affect the patient's well-being, optimism, or despair. Within the construct, a person can interpret an injury or disability to support a positive sense of self or to catalyze self-destructive attitudes. Therefore, one of the more important aspects of the treatment and rehabilitation of patients with disabilities is to be aware that damage and loss of function affect a person's attitude toward self.

## THE EMBODIED SELF

To be one's self is to experience one's body in a certain way. This implies using one's body to carry out intentions in a fluid, unconscious manner. It also means that the body functions as a medium for receiving and conveying information in a fashion appropriate to one's needs and wishes. Thought, desire, perception, expression, and action can be experienced as one seamless piece. Essential to this harmony and embedded quality of psyche and body is the development of what has been called *body schema* or *body image*. The potential for change in this inner percept is critical to our understanding of psychological reactions to disability.

Just as the child develops conceptual schemata of the outside world, so too does it build through sensation and inference an internal image of its body (Schilder, 1950). This is the conscious integration of sensations, perceptions, conceptions, affects, memories, and images from the body surface to its depth. The body schema includes not only our sense of the body within its boundary but the boundary itself and the ways it relates to the environment. The body image plays a vital role in relating our internal experiences, including our drives, to external reality. This image not only facilitates our postures and movements but also anchors our sense of self in a concrete, complexly organized vehicle. Anxiety, drug states, psychoses, or injury can all disorganize and disorient our body image. Body parts that are the source of psychological conflict can drop out of our schema, and the boundary that separates self from environment can be altered during emotionally intense states. Fenichel (1945) noted now body-image distortions were often the earliest forerunner of a later schizophrenic disorganization. Schilder's suggestion that developmental and neurotic factors strongly influence the formation of the body image spurred a continuing interest among psychoanalysts in this topic.

The role of the child's body in developing a sense of identity and in forming a relationship with a differentiated "other" has been stressed by a number of psychoanalytic writers. Greenacre (1958) argued that we reinforce our sense of our body by constant association with others of predominantly similar appearance throughout our life. The development of our body image is always susceptible to distortion and instability. The child exposed to chronic, overwhelming deprivation gives the impression of lacking a sense of body contours and has difficulty in discriminating between "you" and "I." Rank and MacNaughton (1950) emphasized how holding, caressing, and whispering to a child help to create a body image that is endowed with positive, narcissistic value and that ultimately will produce an ego structure with some "strength." They, like other theorists, argued that the personal core, the individuality of the child, must fuse with a body image.

Our most primitive anxieties arise from the fear of disintegration of the body image. Melanie Klein (1975) held that our body image is equated with the ego and finally the self. The child fears losing the entire body; these fears may persist in adults and flare anew with any threat to bodily integrity.

The importance of the body image in facilitating contact with the outside world was articulated by Hartmann et al. (1946), who believed that it is the psychic vehicle for expressing libidinal and aggressive needs. Denial of important parts of the body, particularly the orifices, interferes with ego maturation and ultimately with our relationship to reality. A similar theme was echoed by Winnicott (1953), who spoke of the overgrowth of mental functions that occurs as an attempt to deal with an environment unsatisfactory to the development of the body schema. This overgrowth becomes the basis of a false self that denies its somatic roots.

The ego in the course of life is faced with a changing self, which demands to be synthesized with the abandoned and interrupted selves. This includes changing images of the "body-ego." Erikson (1959) pointed out that our sense of identity and ego include idealized images of our body. The maturation of identity is bound up with the internalized images of our body as it actually exists and the acceptance of its distance from our desired body image. Often as we pass into middle age and beyond, we may neglect our body, sometimes harmfully, because it does not conform to our image of how it should appear.

The "I" is always aware of its self as an embodied "I." This does not imply that the body image is the self but that it reflects the self through its own development. Self, body image, and unconscious striving and attitudes are inextricably related to each other. Schopenhauer (1969) observed that one's body is the objectification of his will. Or in psychological terms, the body and its internalized image is a partial objectification of the unconscious. Each person is shaped by bodily limits and capacities, and in turn, each bears some responsibility for accepting the body and its present abilities.

## THE INTIMACY AND AMBIGUITY OF EMBODIMENT

To be embodied implies a meshing of the psychological and physical aspects of self. We experience an intimate union or belonging to our bodies that makes all other modes of having possible. It often seems that our body is most our own when we are least aware of it, while it performs the simple or complex actions that link our intentions with the world. In illness, however, the intimate union of psyche and soma can be fractured, and the body may come to be apprehended as something foreign to the self. When ill, it is hard to acknowledge that this inert, paralyzed limb or that discolored flesh indeed has any relation to "me." There is, as Richard Zaner (1981) points out, fundamental ambiguity in one's sense of a lived intimacy between one's body and one's self. The body can be experienced as part of the unity of self or as a distanced and distancing object.

Sartre, in *Being and Nothingness* (1956), speaks of the experience of "nausea," the consciousness of our body as an object as revealed by the glance of the Other. A person looking at me may reduce my being to its sheer corporeality, the thinglike aspects of my body and may strip my flesh from my self. My body can be experienced as intimate or mine, but it also has the fundamental property of being experienced as radically other than me. At the very heart of what is most familiar and intimate there is a potential hiatus, an estrangement between my body and my self.

Though our body permits us to be present to the world and achieve certain goals, it also limits the actualizing possibilities of self. We are given and develop a certain body; although we can extend and realize its inherent potentials, there is a horizon to its range of actions and transformations. In our fantasies, as in animated cartoons, we play with the idea of the infinite plasticity of the body, which is always able to reorganize itself to completely express our will. Yet, as Zaner writes, there are simply "some activities, postures, gestures, sensory encounters . . . which are just not within my bodily scope, thanks to my being embodied by this and not some other body" (Zaner, 1981). We dream of having an unlimited scope of use for our own body. The reality is different, although medical science and technology hold out the elusive promise that nature can be extensively transformed. Adult maturation implies coming to accept, explore, extend, and care for the body that we do and will have. The realization of the limits of our body forces us to acknowledge a boundary to our narcissism, exposing us to an uncomfortable awareness of our finitude. Any encounter with the thinglike aspects of our body confronts us with the possibility of the body as inert or dead. It sets up a dread within us, and thus we often deny or distort any change that places a part of the body outside the unity of self.

Any loss of function during illness is frightening. Yet even in health, there are vast domains of body function that lie beyond our ordinary volition. Digestion, breathing, heart rate, and hormone secretions, proceed mostly without our attention or will. These normal processes can create imperatives that constrain and shape the possibilities of volition. What we desire is partly restricted by those rhythms and moods that function automatically. We thus have our body and are constrained by it. But we never complete the task of knowing and being comfortable with our body. It is always in process, changing, making unexpected demands, breaking down, or surpassing our expectations. The body is uncanny because, though it can be alien, it is "ultimately and undeniably mine, expressing and embodying me" (Zaner, 1981).

## POWER AND PLASTICITY IN ILLNESS AND DISABILITY

In discussing intimacy and alienation with respect to embodiment, we observe that to have a body means that at times we will be disappointed in it or that occasionally it will feel alien to us. The very nature of the embodiment of self creates certain problems that may become augmented in an illness. We realize that our body sets limits to the horizon of our activities, and we have doubts that the body we were born with is the one we most desire. We continually experience frustration or dissatisfaction over how it looks or functions. Clearly, if I pined to be a professional basketball center or world-class sprinter, I would consider my body less than adequate. Our somatic development enables all of us to do certain things with our bodies and disables us for others. There is a continuum between health and illness with regard to how I control and value my body. Fears and frustrations lying just beneath the surface in normal periods may erupt and dominate consciousness when I am sick. Even a mild malady such as a toothache can disarticulate the body from the mind. The slightest pain, numbness, or palpitation can suddenly recall my attention from the outside world and focus my concern on some rebellious organ or tissue. I may become flooded with all sorts of worries about my body that may not relate to a current illness but reflect past conflicts and concerns. My sense of competence, identity, status, and power are suddenly called into question when a part of my body escapes my control. If the illness proves transitory, then there are many ways my sense of vulnerability can be shunted out of awareness. Yet to have a chronic illness or disability confronts me with the fragility of my embodiment and threatens to rivet these anxieties to the center of consciousness.

An important aspect of illness and disability is the fear of loss of competence and empowerment. Power is a very important feature of human life. This does not necessarily imply manipulation, influence, or control of the destinies of other people. It does connote, however, the ability of an individual to carry out self-motivated projects in the real world. A person who feels empowered experiences a capacity to initiate and complete a task or goal. To do so requires having the mental, physical, environmental, and interpersonal resources necessary to enact one's wishes. These derive mainly from knowing what one wants, having a plan for enactment, and eliciting the cooperation of important others to implement one's intentions. To have power in this way enhances a person's sense of worth and self-esteem.

Competence refers to the skill with which we execute our projects. An architect may be empowered to design a building, but that does not predict the competence of his execution. There are many factors (which tend to change during a person's lifetime) that affect one's experience of competence and power. Mental abilities, social status, special skills, and talent all play a part. Yet central to each of these attributes is the actual functioning of the body and how it is regarded by its owner and by others.

To be ill or disabled in our society is often perceived by others as a lack of competence, regardless of the person's actual abilities. Patients who suffer some chronic loss of function or disfiguring change in appearance often fear a loss of personal power, worrying that they will lose the ability to influence others and receive in turn a desired response. Another concern centers on the loss of regard by others for themselves as competent and valuable partners, friends, and colleagues. This is especially problematic in our technological society, which both idolizes and discounts the body. To lose validation of one's competence and power through an illness leads one to feel useless and lacking in worth. These feelings are accentuated by the dependency that initially comes during an illness resulting in a disability. We all experience regressive tugs during any period of physical malfunctioning. Our fears about regressing to the status of a child are activated in illness along with latent concerns about our body and its boundaries. The long childhood struggle to gain autonomy of body use is called into question at times of incapacity.

Any loss or impairment of bodily function initially brings with it feelings of denial, protest, anger, and apprehension. The patient worries about the impact the impairment will have on a range of accustomed activities. The person is also concerned about how others will judge him or her. Fears of loss of competence, power, and of gratifying personal relations emerge, which may either lead to social withdrawal or to a new sense of commitment to self-integration. We have noted that the body can be enfolded into the sense of self or disarticulated as an object set apart from "me." There comes a point in any illness or injury when treatment assists patients in reconstituting their sense of body and restoring the experience of wholeness. In doing so, therapists must be aware that patients regard their bodies both from the perspective of their own psychological development and from the vantage of their society.

There is in all cultures a "cognitive consensus", in Durkheim's (1949) sense, about the human body. Each society develops a certain set of beliefs and attitudes toward the body. As Polhemus (1975) points out, there is a communal knowledge of how the healthy, beautiful, or erotic body is defined by a society or subgroups within it. Our bodies become bearers of social meaning within our families and communities. This meaning becomes evident early in childhood and is reinforced through the various stages of development. The education and induction of children into society is inaugurated through the body. It is "the fundamental and universal guide by which human beings understand that they are social beings" (Polhemus, 1975).

How we perceive our physical existence is always filtered through certain social categories. Mary Douglas (1973) tells us that the body communicates information for and from the social system of which it is a part. There is a human motivation to achieve consonance between the social and physical aspects of one's bodily experience. This reconciliation is at best difficult for most people; any impairment of body structure or function creates additional problems, which must be addressed. Family, friends, and colleagues play important roles in any rehabilitation plan in which the patients' concerns over the social perception of their bodies become an important issue. These intimates are vital in helping patients accept their embodied selves and use them as a channel of action and communication. To experience the wholeness of self permits the individual to feel animatedly engaged with the world, to be vividly present to the people and things in his or her environment.

The self can surmount the loss of certain body functions. The embodied self is always possible but must continually be achieved, maintained, and expanded. Working with brain-injured patients, Kurt Goldstein demonstrated how humans strive for optimal performance in the face of organ damage. They attempt to maintain their performance capacity at the highest possible level compared with their former functioning. "When one performance field is disturbed, the most important performances of that field survive the longest and tend to be most readily restored" (Goldstein, 1939). Those performances that survive are the most necessary with regard to the functioning of the whole organism. There is a plasticity of the self that is always seeking to transcend impairment and mobilize resources for self-maintenance and growth. All living systems are endowed to some degree with the remarkable faculty of adaptation. This can refer to the replacement of a lost structure or function by the recombination of remaining elements or to the creation of altogether new types of performance. An amphibian can regenerate a lost limb, and a bacterium can produce, when taken from one culture medium to another, an enzyme never before found in its metabolic repertoire. Humans do not demonstrate this kind of structural plasticity, but we do have the capacity to restore a unity of mind and body, which expresses itself in the presentation and engagement of the self to the world.

The papers in this volume speak to the endurance and thrust of the self in the face of congenital or acquired bodily impairment. Implicit among all the therapeutic suggestions touched on by the authors is the need to help the patient regard the body as an instrument capable of effective action. Wherever denial, guilt, shame, or anxiety about the body occurs, it should be acknowledged and understood by clinicians. Disability is no barrier to the development of the self. We all struggle to reidentify with our body as it passes through childhood, adolescence, adulthood, and late maturity. The self develops as it experiences an integration of its faculties and a continuing ability to express new competencies. Each of the contributors affirm in his or her way that where an embodied self exists, there are always possibilities.

## SUMMARY

In this essay I have suggested that the clinician regard the body in two different ways—from a scientific perspective and as an indissoluble aspect of self. We must of course understand any illness in a careful, scientific manner by bringing to bear on its diagnosis and treatment all of our knowledge gleaned from anatomy, physiology, and cell biology. The mechanisms of pathology and therapeutics must be understood and used when the clinician accepts responsibility for the care of a patient. And yet the body is more than a collection of cogs, drains, and chemical engines. As de Lignac (1939) has stated, spirit and matter are always present together in human activity, the two elements bonded in a taut relationship. To betray either the body or the psyche is to invite the revenge of both, and the payment is sickness. If the mind neglects or humiliates the body, it loses its balance and measure. Psychic life itself is disturbed by repression of the body: when threatened with neurosis, the psyche loses touch with reality. We err in both overmechanizing the body and in forgetting it in the glorification of the soul. Often our apprehension of death and of our basic impulses leads us to ignore, mythologize, or fear our bodies. The poet and philosopher Octavio Paz (1974) wrote that "the body is imaginary, not because it lacks reality, but because it is the most real reality—an image that is palpable yet ever-changing, and doomed to disappear." St. Teresa of Avila recognized our biological rootedness when she said, "We are not angels, for we have a body."

As children we learn about suffering, frustration, and our limitations through the body. We are not necessarily pleased by the discovery of our independent corporeal existence, because it forecasts the dangers of isolation, the inadequacy of our defenses, and the obstacles to satisfying our needs. But it is also through the body that we can make contact with our self, with our unique, evolving nature. This experience has been beautifully described by the Czech novelist Milan Kundera (1984):

Tereza tried to see herself through her body. That is why, from girlhood on, she would stand before the mirror so often. And because she was afraid her mother would catch her at it, every peek into the mirror had a tinge of secret vice.

It was not vanity that drew her to the mirror; it was amazement at seeing her own "I." She forgot she was looking at the instrument panel of her body mechanism; she thought she saw her soul shining through the features of her face. She forgot that the nose was merely the nozzle of a hose that took oxygen to the lungs; she saw it as the true expression of her nature. (Kundera, 1984, 41)

In this paper I have emphasized that successful rehabilitative work with disabled persons requires clinical appreciation of the meaning of the body for an individual. Yet it is often as difficult for patients to talk about their bodies as it is for clinicians to listen. Clinicians must, however, find ways to gently help patients reveal to themselves and discuss their fears, hopes, guilt, and pride about their bodies. And clinicians must become more aware of their own myths and misgivings. Patients are uncomfortable having their damaged bodies observed, manipulated, and evaluated (even for treatment) by others. Fear of losing mastery or control over one's body is a threat to the experience of wholeness of self. Our sense of personal unity is forged from our daily struggle to differentiate and integrate our physical and mental capacities. It has been said that "thought must be taught to dance." A clinician's task is to help patients discover, whatever their loss, which dance is possible for them.

## REFERENCES

Allport, G. (1955). *Becoming.* New Haven: Yale University Press, 36–56.

Critchley, M. (1955). Personification of paralyzed limbs in hemiplegics. *British Medical Journal, 2,* 284.

de Lignac, J. A. L. (1939). The relationship of sexuality with the person. *Body and Spirit.* London: Longmans, Green, 29.

Douglas, M. (1973). *Natural symbols: Explorations in cosmology.* New York: Vintage Books.

Durkheim, E. (1949). *The division of labor in society* (trans. G. Simpson). New York: Free Press. (Orig. pub. 1893).

Erikson, E. (1959). The problem of ego identity. *Psychological Issues, 1,* 149.

Fenichel, O. (1945). *The psychoanalytic theory of neurosis.* New York: W. W. Norton.

Freud, S. (1964). *The ego and the id.* New York: W. W. Norton, 16.

Greenacre, P. (1958). Early physical determinants in the development of the sense of identity. *Journal of the American Psychoanalytic Association, 6,* 612–627.

Goldstein, K. (1939). *The organism.* New York: American Book, 47.

Hartmann, H., Kris, E., & Loewenstein, R. (1946). Comments on the formation of psychic structure. *The Psychoanalytic Study of the Child, 2,* 11–38.

Hume, D. (1739). *An abstract of a treatise of human nature.* Cambridge, England: University Press.

Idler, E. (1979). Definitions of health

and illness and medical sociology. *Social Science and Medicine, 13A,* 723–731.

Klein, M. (1975). On the theory of anxiety and guilt. *Envy, gratitude and other works.* London: Hogarth, 25–42.

Kundera, M. (1984). *The unbearable lightness of being.* New York: Harper Colophon Books.

Merleau-Ponty, M. (1964). Phenomenology and the sciences of man. In J. Wild (Ed.), *The primacy of perception.* Evanston, Illinois: Northwestern University Press, 5–83.

Paz, O. (1974). *Conjunctions and disjunctions.* London: Wildwood House, 14.

Polhemus, T. (1975). Social bodies. In J. Benthall & T. Polhemus (Eds.), *The body as a medium of expression.* New York: E. P. Dutton, 13–35.

Rank, B. & MacNaughton, D. (1950). A clinical contribution to early ego development. *The Psychoanalytic Study of the Child, 5,* 53–65.

Reich, W. (1949). *Character analysis* (3d ed.). New York: Orgone Institute Press.

Sartre, J. P. (1956). *Being and nothingness.* New York: Philosophical Library.

Schilder, P. (1950). *The image and appearance of the human body.* New York: International Universities Press.

Schopenhauer, A. (1969). *The world as will and representation.* New York: Dover.

Winnicott, D. W. (1953). Transitional objects and transitional phenomena. *International Journal of Psychoanalysis, 34,* 89–97.

Zaner, R. N. (1981). *The context of self.* Athens, Ohio: Ohio University Press.

Zegans, L. S. (1982). The self and systems theory. In W. Gray, J. Fidler & J. Battista (Eds.), *General systems theory and the psychological sciences, 2,* 11–23.

Zegans, L. S. & Victor, B. (1985). Conceptual issues in the history of psychiatry. In H. H. Goldman (Ed.), *Review of general psychiatry.* Los Altos, California: Lange Medical Publications, 5–22.

# 2

# Preparing for Uncertainty: Family Reactions to a Seriously Impaired Child

*Albert J. Solnit*

The title conveys one contradiction and one ambiguity. How can we prepare for uncertainty when the concept of uncertainty contradicts the thrust of preparing? Does a family react as a unit, or does each member, each parent or sibling, have at least individual reactions that, while interactive, remain embedded in the unique psychology of each individual? What follows will not overlook these "noises" but address them with the aim of showing how they can be transformed into background "music."

Children are born helpless. Without the care of a committed, affectionate adult, they do not survive. Recently in London, a young unwed mother died suddenly at home of what appeared to be a rupture of a congenital aneurysm. Her healthy infant child died of starvation because no one knew the mother had died.

When a child is born, he or she enters into a world prepared to care for a healthy, intact, and helpless child. By their responses and development, healthy newborn children reassure their families (i.e., parents, siblings, grandparents, and others) that they are biologically intact. The child's growth and development can steadily provide evidence of parents' competence and of the strength of their hereditary contribution, implying a promising future. Usually, the community is prepared to provide the support and services necessary for a healthy newborn child.

## REACTIONS TO IMPAIRMENT

When a young child is developmentally disabled or has a congenital or inherited defect, the parents feel injured. The sad, resentful, discouraged reactions of the parents, which have been referred to as mourning reactions for the healthy child that was expected (Solnit & Stark, 1961), require that time and support be given to the parents to enable them to prepare for the care of the child they did not expect. Such time and support is often not available. This tends to start a downhill spiral, in which the average environment, one not prepared for the disabled child, becomes an environment of risk for that child and its parents.

At the same time, sibs feel threatened by uncertainty, by the further withdrawal of their parents' attention, and by the fear that their own future physical and social well-being is jeopardized. In the newborn period, these challenges usually find the mother and father, but especially the mother, particularly vulnerable to what may be felt as catastrophe. Thus, there is a good deal of reason to assume that the physical or functional disability of the child tends to be elaborated by the initial responses of parents and other members of the family, who have been traumatized by the birth or development of a deviant child. Defects evoke weakness before they can evoke strength.

A similar case can be made in regard to many physicians, nurses, social workers, and others who feel threatened by the birth or development of a defective child. They are likely to feel that their competence has been undermined by the failure of the child to be born and grow in a healthy manner. Additionally, they often feel discomforted by their inability to make precise forecasts, or to provide clear explanations of what has happened to bring about the condition.

In general, obstacles to the care of the developmentally disabled child are related to the dynamic equation that health brings acceptance and support, whereas disability evokes an increasing health risk in an environment that was organized for the able child. One of the major expressions of these dynamic relationships is the tendency for a family with healthy, resilient relationships to mobilize its resources to master the challenge, whereas a troubled, underprivileged, or disadvantaged family tends to become all the more unable to cope when a defective or disabled child becomes a member of their family. The planned study of these obstacles will enable us to become increasingly specific in understanding how defects evoke weakness and how health evokes confidence and competence.

## PARENTS AND PROFESSIONALS

Professionals are more likely to righteously impose services in the care of a disabled child if they feel the family has not responded as the physician, nurse, or social worker feels they should. Thus the defensive reactions of professionals tend to discourage a family's self-help attitude and interfere with the development of a therapeutic alliance between the family and the health workers. That is, while the family is still absorbing the experience of having a disabled child, the professionals often respond initially to their own urge to master the challenge of the child's disability and are not in touch with the family's mourning and coping reactions. The tempo and rhythm of parents' and professionals' responses, of becoming active in helping the child, are sufficiently different to interfere with the collaborative activities of these adults. The professionals cannot make their knowledge and skills useful without the knowledge and skills of the parents; conversely, the parents are less effective and satisfied in helping their child if they do not have a mutually respectful and sustained partnership with specific professionals.

Added to this, the complexity and scarcity of multidisciplinary services make it very difficult for parents to discover how to enter the system of services that are available, thus leaving such parents with a magnified sense of helplessness and hopelessness. Siblings sense this negative perturbation and, according to their ages and their unique histories, respond with bewilderment, feelings of abandonment, feelings of guilt, and/or reactions designed to cope with and master these emotional and cognitive stresses. Regarding services, paradoxically, the more we know and the more skills we have in our repertoire, the more complex these programs and systems become.

Finally, our limitations in organizing and coordinating multidisciplinary services are a painful indicator that our ability to treat specific aspects of a disability has far outstripped our capacity to orchestrate diagnostic and therapeutic rehabilitation and supportive services. Yet we know these should be available for disabled children and their families. It is this final point that impelled Dr. Ronald MacKeith to write, under the title "The Buck Stops": "The care of the [seriously impaired child] goes on for a long time and is usually complex, at first medically and later educationally and then vocationally, with medical, psychiatric, and social issues playing larger or smaller parts at various periods. . . . Some one person has to feel responsible and the various agents involved have to accept him as their coordinator" (MacKeith, 1969, 691–693). At any time such a coordinator and health professional ombudsman can enhance the confidence and authority of the parents in their care of the impaired child as well as his or her sibling.

Parents should be free to reject or to follow the advice offered. If they decide to ask experts, they should be free to decide whether and how to follow their suggestions. This freedom is the basis for confidentiality and for the capacity to actively help oneself and one's child. Such freedom implies that the alliance between parents and professionals is a true collaboration in which each needs the other but is also free (without threat of punishment or retribution) to question and to disagree with the other.

In one dramatic instance, an autistic psychotic child, age 6, engaged in violent, self-mutilating behavior and at other times attacked household materials and, occasionally, his parents and siblings. The parents understood this as the behavior of a sick child. They tried to curb it through a well-structured and simplified environment, through the elimination of environmental hazards, and through the use of professionally advised psychological and pharmacological treatment.

They were able to mitigate all but the violent and self-destructive behavior in which the child cut himself with any piece of glass he could find and break to form a sharp or pointed edge. The mother described how the child could detect pieces of glass in the yard that had been overlooked by the family's extensive effort to eliminate the hazard. He could detect it by his extraordinary sensitivity to bright, flickering light, which such pieces of glass reflected when the sun was out. The parents refused the child psychiatrist's advice to hospitalize their child.

Instead they turned to the guidance of the National Society for Autistic Children. They were advised to follow a regime of behavior modification, which carefully delineates the way to use painful, aversive conditioning as a teaching method—not a treatment—to rapidly bring under control behavior that "threatens the child's safety or his survival in an optimum environment. Such aversive conditioning may involve spanking and electric shock."[1] Parental autonomy and the use of physically painful conditioning can be viewed as efforts to maintain the integrity of the family or can be seen as a problem of child abuse. One could say that the question of physical abuse in childhood should take into account both the complexity of what is involved and especially the available alternatives. The use of aversive conditioning under these circumstances was viewed as the least detrimental alternative in enabling the parents to maintain the integrity of their family while protecting their child's physical safety, that is, curbing his self-destructive behavior (Goldstein et al., 1979).

## LIFE OR DEATH

When a child suffers from a chronic, life-threatening illness, which even in the recent past has been viewed as fatal, the family and the child often find themselves in uncharted waters (Green, 1980). Children suffering from leukemia, lymphomas, sarcomas, and other malignancies, as well as certain inborn metabolic disorders (e.g., cystic fibrosis), are often in a situation in which adults, parents, and professionals cannot provide firm guidance based on confident expectations. The adults and children do not know whether to prepare for life or death.

Throughout, ambiguities, especially about preparing to die or preparing to live, will pose a challenge to those in the healing arts and to the child and its family. Presumptively, the least harmful attitude for those providing the medical care is that of cautious, reserved optimism (Tiger, 1979). Such optimism is as realistic as restrained pessimism, since each child is unique and not to be treated as a statistical mean. This attitude acknowledges that our precision in making forecasts is limited. It also recognizes that where such limitations are in force, the hopes for and commitments to a life worth living can be sensibly given the weight of a feather on the side of optimism. Any more than a feather's weight would be like a promise and broken promises are damaging and distressing. This slight emphasis also supports the continuing concern of professionals in the care of their patients and enables patients to better feel the supportive presence of caretakers. In contrast, if professionals feel hopeless or resigned before it is completely clear that there is no hope, patients and their parents usually perceive this as withdrawal, apathy, or insensitivity.

Is such an approach an either/or trap, one that comforts those who need it least (doctors and nurses) and deprives those who need it most (the child and its parents) of an uncertainty, an ambiguity that is painful but realistic? Can one not know and be neither optimistic or pessimistic? Will a preference for a slight optimism corrupt those who should lead with knowledge and humility and disillusion the patient and the parents?

We all wish for certainty, knowing it is rare. When certainty is apparent, most frequently there follows a cynical disillusionment when it turns out to be illusive. We are struggling with the painful dilemma that knowledge advances slowly, piece by piece, and our ability to put it to use is hampered because we want new part-truths to be whole and absolute. This dilemma is increasingly difficult to avoid as our understanding about chronic and fatal illness in childhood grows at, by comparison to forty years ago, a breakneck pace. How shall professionals guide themselves when parents want to know if the pain and destructive effects of a new treatment and the repetition of illness and treatment are worth the candle for a child who is suffering from an acute lymphoblastic or myeloblastic leukemia, a child with no remaining kidney function, a child with an inborn error of metabolism or a congenital or acquired immunological deficiency? How shall doctors assist parents who are confronted with crises that threaten their child's life over and over and who have a limited ability to provide hour-to-hour care?[2]

## GUIDELINES FOR THE LAW

In fact, psychoanalytic knowledge and its limitations have strongly suggested that such decision making rest with the parents as professionals provide them with as much information and support as they can. On the other hand, parents may prefer to delegate the decision to clinical experts or the courts. We have suggested the following questions as guidelines to indicate the limits of parental autonomy in making such decisions:

**1.** Do medical experts agree that the treatment is nonexperimental and appropriate for the child?

**2.** Will the denial of treatment result in an early death?

**3.** Is the anticipated result of treatment what society would want for every child? Will the treatment lead to a life worth living? (Goldstein et al., 1979)

Parents who would not allow such a treatment when all three questions have been answered affirmatively have justified court intervention to provide such a treatment (under the authority of the state) before returning the child to its parents. Conversely, if any one of the three questions cannot be answered in the affirmative, there would be no grounds on which to interfere with the parents' preferences regarding the provision of treatment. It is the responsibility of professionals to help parents know their options and to support the family in carrying out its decisions (Duff & Campbell, 1979; Lewis & Lewis, 1979; Schowalter, Ferholt & Mann, 1973). If a child suffers from a chronic illness that would constitute a threat to life if untreated, for example, juvenile diabetes melitus or severe repeated grand mal epilepsy, the parents' refusal or inability to provide the daily care and treatment that is necessary should also prompt court intervention to protect the child from being physically damaged.

## DISCUSSION

With these three guidelines, the search for certainty is limited, and coping with uncertainty in regard to invariably fatal conditions (e.g., malignancy, no kidney function, an urgent need for blood transfusion as in aplastic anemia) can be approached rationally. Statistical probabilities shift too rapidly and are too limited for guidance in the individual case to take the rights of parents and their children away with regard to living and dying unless all three guidelines are satisfied.

These issues are also raised to emphasize how complex the decision-making process is with regard to the newer treatments—radiological, surgical, and chemical. We must be prepared to accept our limitations in making predictions. This should also encourage professionals to share their knowledge with parents and to provide support for them as they participate in making decisions about treatment or as they delegate to professionals responsibility for decisions to treat. In this formulation it is assumed that treatment ranges from specific therapy that will cure, with or without toxic and other untoward side effects, to nonspecific therapy that provides appropriate symptomatic relief or as much freedom from pain, discomfort, and fear as possible. It is in this sense that choices of therapy can be presented for decisions by parents, older adolescent patients, or physicians when delegated.

With such uncertainty and with increasing opportunities for parents and children to make choices, how will they know what to expect? How will they know whether to prepare for dying or for a life worth living? Perhaps we can stretch the human capacity to tolerate not knowing if we learn what to do about it and how to do it. Physical pain, loneliness, helplessness, and anxiety are still basic discomforts, each of which threatens to exceed the child's tolerance in damaging ways. Thus symptomatic relief is crucial. In addition, enabling the patient and parents to know, at the level they want to know, can be relieving for them. Forming alliances with the children and their parents that enable them to have an active role in the decisions about diagnostic and therapeutic procedures at the same time as the clinical expert is respectful of their preferences, for denial can be supportive (Duff & Campbell, 1973). For parents to meet and help each other in group therapy and group discussions has been reported as helpful (Adams, 1978; Gilder et al., 1978). Coping with similar fears and uncertainties, these parents, whose children are involved in similar painful and distressing treatments and uncertainties, share the same hopes that their children, like Lazarus, will return to the living. However, such group experiences are not useful to everyone. They should be viewed as an option for parents to consider and to accept or reject as a preferred pathway of assistance.

Above all, the mantle that physicians and other health professionals have been invited to wear for so many years must be resisted and modified, especially those coverings that suggest fantasies of omniscience and omnipotence. Competence, Compassion, and Continuity of care are the key requirements for health professionals who are working together as a team with seriously ill children and their parents, who are facing an uncertain outcome. These three Cs will be eroded if certain principles are not followed: First, one member of the health team is the physician of record (Duff, 1979), the one who provides continuity and orchestrates the medical care to assure competence with compassion; second, the recognition of the limitations of our knowledge, technology, and inner resources (Schowalter, 1978) when working with children who have what may be a fatal disease. As one colleague put it:

Passing through a hospital playroom, a staff social worker paused near Dana, who, when he was five, had become blind from an inoperable brain tumor. Feeling a surge of sadness about this little boy's predicament, wishing to reach out and offer comfort, the social worker nonetheless hesitated awkwardly—what can you say to such a child? How can you talk to him?

Sensing her nearness, Dana turned toward her and held out his hand. 'Would you help me go across the room to where the blocks are, please. I can't see, you know. I don't want to bump into something.' His quiet explanation about his condition made possible a comfortable and spontaneous conversation. Child and social worker found their way across the room together, talking about the block building that Dana was planning. (McCollum, 1981, 71)

I was that social worker. And late that afternoon, my family came home to find me seasoning the supper casserole with tears—tears shed for Dana.

A clinical social worker becomes a finely tuned instrument—a receptive instrument that registers and resonates with nuances of her client's feelings. We call it empathy, without which there can be little, if any, therapeutic gain. At times, however, empathy can merge into identification with patient or client. When that occurs, the sense of separateness needed for a therapeutic perspective is lost. The client may be less effectively served, and the clinician may feel unduly burdened. (McCollum, cited in Solnit, 1983)

In fact, those who go beyond the limits of their own inner resources often become those who invest in either/or approaches, insisting on unreal certainties that protect themselves but often expose the sick child and parents to oversimplifications, categorical imperatives, and formulae that tend to dehumanize the decisions and the care for the child.

As we have become more familiar with openness toward chronic illness, serious impairment, and dying, there have been many efforts to reduce the dehumanizing aspects of multiple specialists involved in treating a child suffering from a fatal disease. Once the treatment can be carried out on an ambulatory basis, for example, families can be given their choice of where their child will die, in hospital or at home. In a remarkable study (Fortunato & Komp, 1979), it was demonstrated that children who died at home could receive the medical help and support they needed, that the families had made this choice freely rather than hospitalizing their dying children, and that those 13 of 27 who died at home (median age, 5½ years) were comparable in terms of illness, treatment, and distance from the hospital.[3] They conclude, "The home environment can contribute much to the supportive care and comfort of a dying child. The long period of time from diagnosis to death in *all* can permit the family with the proper guidance to develop a more complete acceptance of the child's death" (Fortunato & Komp, 1979, 124).

## SUMMARY

In this paper I have examined the developmental challenges and risks for parents and siblings and the role and limitations of health professionals when a child is under treatment for a disease that poses a serious, uncertain threat to life. The examination of these situations, in light of comparatively rapid changes in our knowledge of life and death issues for such children, points to the complications that arise when certainty is insisted upon by parents and by those who provide professional care and guidance for the sick child and its family. These complications deserve continuing, systematic study as a first step toward reducing and preventing them, and as a means toward a better grasp of the painful question, do we prepare for life or death?

## NOTES

1. White Paper on Behavior Modification with Autistic Children, revision by Clara and David Park and members of the Professional Advisory Board,

National Society for Autistic Children, of a statement drawn up by Creighton Newsome (6/12/75).

2. According to the family's values and in agreement with societal consensus.

3. Candor and tact about diagnosis, prognosis, and treatment often foster innovation by parents and health professionals. Thus, more home care may be desirable and feasible.

## REFERENCES

Adams, M. (1978). Helping the parents of children with malignancy. *Journal of Pediatrics, 93,* 734–738.

Duff, R. S. (1979). Guidelines for deciding care of critically ill or dying patients. *Pediatrics, 64,* 17–23.

Duff, R. S. & Campbell, A. G. M. (1973). Moral and ethical dilemmas in the special-care nursery. *New England Journal of Medicine, 289,* 890–894.

Duff, R. S. & Campbell, A. G. M. (1979). Social perspectives on medical decisions relating to life and death. In J. Ladd (Ed.), *Ethical issues relating to life and death.* New York: Oxford University Press.

Fortunato, R. & Komp, D. (1979). An evaluation of death at home for children with acute lymphoblastic leukemia. *Virginia Medical Monthly, 106,* 124.

Gilder, R., Buschman, P. R., Sitars, A. L. & Wolff, J. A. (1978). Group therapy with parents of children with leukemia. *American Journal of Psychiatry, 32,* 276-287.

Goldstein, J., Freud, A. F. & Solnit, A. J. (1979). *Before the best interests of the child* (new ed. with epilogue). New York: Free Press.

Green, M. (1980). The "vulnerable child"—Intimations of mortality. *Pediatrics, 65*(5), 1042–1043.

Lewis, M. & Lewis, D. (1979). Death and dying in children and their families. *Archives of the Foundation of Thanatology, 7,* 9.

MacKeith, R. (1969). The buck stops. *Developmental Medicine & Child Neurology, 2*(6), 691–693.

McCollum, A. T. (1981). *The chronically ill child—A guide for parents and professionals* (rev. and enl. ed.). New Haven: Yale University Press.

Schowalter, J. (1978). The reactions of caregivers dealing with fatally ill children and their families. In O. J. Sahler (Ed.), *The child and death.* St. Louis: C. V. Mosby.

Schowalter, J., Ferholt, J. B. & Mann, N. M. (1973). The adolescent patient's decision to die. *Pediatrics, 51,* 97–103.

Solnit, A. J. (1983). Changing perspectives: Preparing for life or death. In J. Schowalter, P. R. Patterson, M. Tollmer, S. V. Gulbo & D. Peretz (Eds.), *The child and death.* New York: Columbia University Press.

Solnit, A. J. & Stark, M. H. (1961). Mourning and the birth of a defective child. *The Psychoanalytic Study of the Child, 15,* 523–557.

Tiger, L. (1979). *The biology of hope.* New York: Simon & Schuster.

# PART II

## Assessment

The key to effective treatment and successful rehabilitation is careful, competent assessment. Such an assessment should include physical, social/vocational, psychological, and neuropsychological evaluations. Clearly, this effort requires the skills of multiple specialists. The chapters in this section discuss issues in the psychosocial and neuropsychological evaluation of persons with physical disabilities.

Mitchell Rosenthal describes the comprehensive rehabilitation of persons with physical disabilities as a holistic enterprise in which professionals, family members, and the person suffering the disability work in concert toward identical goals: to achieve as much independence and as high a quality of life as possible. Rosenthal discusses some common but mistaken assumptions about disability. He suggests that long-term physical disability rarely causes severe psychiatric disability, that reactions to disability do not proceed through a standard series of stages, that the adaptive process is not time limited, that depressive reactions are not necessarily maladaptive, that professional intervention is not always necessary for successful rehabilitation, and that the degree of maladaptation often does not correlate with the degree of functional impairment.

Optimal treatment and rehabilitation require that psychosocial assessment be considered a primary rather than a secondary task. Rosenthal reviews the multiple sources and types of data the psychologist uses to construct an impression and contribute to the team's formulation of a treatment plan. Sources of data include medical records; educational and vocational reports; observations and history from the individual (and family); clinical interviews; psychological, neuropsychological, educational, and vocational testing; and finally, reports of teachers, fellow workers, and community agencies, which can often provide important longitudinal information. A good assessment should yield an understanding of the person's current psychological and physical strengths and limitations; premorbid history; level of economic resources; recreational and avocational interests; and behavioral, emotional, cognitive, and sexual functioning. An appraisal should be made of past patterns of behavior, coping strategies, quality of interpersonal and family relationships, and the implications of these factors for successful rehabilitation. After describing various kinds of assessment methods, Rosenthal uses case vignettes to present some common behavioral problems and suggests useful interventions.

In the past twenty years, neuropsychological assessment has for two major reasons assumed increasing importance in the evaluation of persons with physical disabilities. First, because of a growing awareness of the prevalence of neurological impairments associated with physical disabilities and second, because neuropsychological assessments have proven reliable, valid, and useful in understanding the patient's status and in guiding any interventions.

Ralph Kiernan begins his chapter on the differentiation and assessment of physical, cognitive, and emotional disabilities by discussing the lack of meaning of the global term "organicity" and the assessment instruments developed to evaluate specific deficits in neuropsychological functioning. Kiernan describes the necessity for sensitive as well as insensitive tests of functioning, for instruments of broad scope as well as fine-grained focus. He encourages the examiner to develop, test, and modify specific hypotheses, rather than merely administering a standard battery. Especially in cases of physical and emotional disability, it is essential to choose tests which allow the patient the least restrictive avenue for performance. This may be difficult enough when contending with the impact of physical impairment on test performance; the overlay of disabling emotional factors often makes the task even more challenging, taxing the assessor's ingenuity. For Kiernan, meeting this challenge means abandoning a cookbook approach to assessment and tailoring each evaluation to the individual patient to overcome, insofar as possible, any limitations in the patient's physical abilities or complexities in presentation. He offers a series of guidelines, strategies, and case examples to illustrate his tailor-made approach to the assessment of physical, cognitive, and emotional factors in persons with sensory and physical disabilities.

# 3
# Psychosocial Evaluation of Physically Disabled Persons

*Mitchell Rosenthal*

Comprehensive rehabilitation of the physically disabled is a holistic enterprise. The modern rehabilitation center, whether in an acute hospital or a free-standing institution, is staffed by a variety of medical, health care, and psychosocial specialists—including physicians; physical, occupational, and speech therapists; rehabilitation nurses; rehabilitation psychologists and neuropsychologists; social workers; vocational counselors; recreational therapists; and dietitians. Their goals are complementary and often identical: to assist the physically disabled individual and family regain as much independence as possible in order to achieve a high quality of life, regardless of the specific nature of physical impairments that accompany a given disability. For rehabilitation to be successful, psychosocial assessment and intervention must be considered a primary, not a secondary, task for the rehabilitation team.

The aims of this chapter are to (1) describe some central concepts, which define the process of adaptation to physical disability, (2) detail the source and types of assessment data, (3) outline some methods of assessment, (4) describe the process of synthesizing such information and translating it into treatment strategies, (5) review some characteristic psychosocial problems in the physically disabled patient, and (6) illustrate through several case examples some of the principles and practices that have been described.

Though this chapter is written from the perspective of a rehabilitation psychologist, the procedures described are often performed by social workers, psychiatrists, rehabilitation counselors, and others, depending upon the specific setting and the expertise of the professionals there. Psychosocial assessment should not be the sole province of psychosocial team members, but rather a mandatory area of investigation for all members of the rehabilitation team.

## COMMON ASSUMPTIONS ABOUT DISABILITY

It is often assumed that a physical disability is a catastrophic event that in turn creates severe emotional disturbance. In fact, though onset of a disability may create extreme stress and often result in great emotional distress, it is rare that physical disability causes long-term psychiatric disability.

Another common belief is that individuals experience identical reactions to disability and progress in a logical manner through a series of stages. Until fairly recently, the common wisdom was that physical disability causes an individual to progress through stages similar to a person facing terminal illness (e.g., denial, depression, bargaining, and acceptance). Trieschmann (1980), in discussing the psychological, social, and vocational aspects of spinal cord injury, effectively countered these arguments by a critical examination of the research literature and concluded that evidence was lacking for a universal reaction or a set of invariant stages in response to disability.

It has often been stated that the adaptation process is time limited. Within a period of six months, for example, most people have fully accepted and adapted to their disability. In fact, any clinician who has followed spinal cord– or brain-injured patients for several years will readily agree that the process of adjustment is continual and may be lifelong, for the patient as well as the family of the disabled person.

Another myth holds that depressive reactions are maladaptive. If one visits a typical rehabilitation unit, it is easy to make the clinical observation that most patients appear depressed. Many are quite sad, teary, and voice pessimism about the future. These are appropriate reactions to a catastrophic life event. To the rehabilitation professional, depression signals an understanding of the nature of the disability. It is only when the depressed behavior interferes with the individual's participation in an active therapy program and is reflected in social withdrawal that depression may be considered maladaptive.

To adapt successfully to disability, one must rely on professional intervention. This assumption falls with the frequent observation that individuals are able to adapt through their own resourcefulness, with the help and support of family or friends, or through peer counseling, that is, emotional support and guidance from another person with a similar disability. Peer counseling is sometimes provided through a local independent living center (ILC) or through established disability associations (e.g., stroke clubs, Easter Seal Society, family support groups, National Head Injury Foundation).

Yet another fiction is that there is a correlation between the severity of disability and the degree of maladjustment. For example, a below-knee amputee can adjust better than an above-knee amputee; a paraplegic can adapt better than a quadriplegic; a patient with a minor concussion can cope better than one with a severe head injury. In fact, there is very little correlation. Coping or successful adaptation is contingent upon a variety of complex personal, environmental, and physiologic variables, which vary from person to person. These variables will be described later in this chapter.

## SOME TRUTHS ABOUT DISABILITY

The presence of a disability may be viewed as a stressful life event. Selye (1974, 14–15) defined stress as the "nonspecific response of the body to any demand upon it." The stress may be acute, in which threat is immediate and response must be instantaneous, or chronic, in which both stress and response are prolonged and unabated (Pelletier, 1977). For the disabled, stress is most often both acute and chronic and thus requires continual adjustment and readaptation. Common stressors facing the physically disabled are the initial diagnosis and prognosis, surgery, transfer from one hospital unit to another, impending discharge, and reintegration into the family and community, among others.

Almost every sphere of life may be affected by the presence of a physical disability. Loss of mobility function, dependence in activities of daily living, altered bodily functions, loss of income, change in living arrangement, altered peer and family relationships, decreased ability to engage in educational and vocational pursuits, and diminished capacity to engage in recreational pursuits are but a few of the areas that are affected.

Psychosocial adaptation is a function of an individual's premorbid behavior patterns as well as physiologic and environmental variables. Though it cannot be denied that the degree of physical impairment has some relationship to coping ability, it is clearly not the only factor. This conceptual framework has been advanced by Trieschmann and others and is illustrated in figure 3.1, which lists some of the key variables as they apply to spinal

**P = Person Variables**       **O = Organic Variables**

Repertoire of habits      Age
Personality style      Severity of disability
Rewards and preferences      Medical complications
Internal vs. external locus of control      Congenital anomalies
Method of coping with stress      Strength
Creativity      Endurance

$B = f(P \times O \times E)$. Behavior (B) as a function of the Person (P), the Organism (O), and the Environment (E).

**E = Environmental Variables**

Hospital milieu
Stigma value of disability
Family and interpersonal support
Financial security
Social milieu
Urban vs. rural residence
Access to medical attention and equipment repair
Access to educational, recreational and avocational pursuits
Socioeconomic status
Architectural barriers and availability of transportation
Legislation
Cultural and ethnic influences

**FIG. 3.1.**

*Source:* R. Trieschmann, *Spinal cord injuries: psychological, social and vocational adjustment* (New York: Pergamon, 1980), with permission.

cord injury. The work of DeJong (1979) on the independent living outcomes for severely disabled individuals has highlighted the crucial role of environmental variables in determining long-term outcome.

Disability has a significant impact upon family members, who can greatly influence the rehabilitation process. English (1983), in reviewing the literature on the family's role in rehabilitation, concludes that "the most global theme is that one person's disability has a profound and lasting impact on nuclear family units and often on extended family systems as well. . . .disabled person's families are disabled as well and often in need of assistance" (9). Versluys (1980) points out that family response to disability may be positive, characterized by emotional support, constructive planning, and the like, or it may be negative and pathological, characterized by overprotection, reinforcement of dependency behavior, neglect, avoidance of discharge planning, denial of the diagnosis or prognosis, inappropriate demands on staff or relatives, or open rejection. These potential constellations of behaviors suggest how the rehabilitation process can be affected by the behavior of family members.

## THE PSYCHOSOCIAL ASSESSMENT

### Sources of Data

The psychosocial assessment team collects data. These data comprise a variety of types of information obtained from a variety of sources. First, the medical record is an important source of data, which provides a good summary of pertinent medical findings, current functional status, the rehabilitation plan, and the estimated length of stay. The rehabilitation physician's initial history and physical usually contain a brief summary of the patient's previous living arrangement, family constellation, and relevant educational or work history. In the case of children and young adults, data from educational records can provide a very important baseline that can be used to assess disability-related changes in cognitive function.

The next step in the process is the clinical interview. The clinical interviewer tries to obtain an in-depth history of the patient prior to the current hospitalization (often prior to the onset of the disability) and an appraisal of the patient's current understanding of his or her disability and the goals for further rehabilitation. In the case of neurologic impairment, for example, the patient may be an unreliable informant; it is therefore critical to interview the family. When the patient is a young child or preadolescent, a family interview is essential. A comprehensive picture of the patient's lifestyle, personality, and behavior patterns should emerge from the clinical interview.

Often, the psychosocial assessment is the basis of future intervention strategies. Patients with disabilities often display maladaptive behavior on the unit or in therapy. Though sometimes these behaviors can be elicited within the clinical interview, more often, they need to be viewed in their natural environment—the ward, rehabilitation clinic, physical therapy gym, school, or work place. Careful behavioral observation will lead to a better formulation of intervention strategies.

Psychological, educational, and vocational testing is another important part of the overall assessment. Often, referral questions from physicians specifically focus on the level of intellectual functioning, nature of residual neurobehavioral deficits, prognosis, personality/behavioral profile, likelihood of returning to similar educational or vocational activities, or need for retraining for certain skills or abilities. Within the context of the total assessment, psychological testing is often an important tool by which such information can be obtained. The areas of assessment usually include cognitive abilities and disabilities, social-emotional competence, academic skills, perceptual-motor capacities, vocational interest, and the like (Guidubaldi et al., 1979). Neuropsychological assessment, a specialized form of testing that examines brain-behavior relationships, is often performed with patients having evidence of neurologic disability: head trauma, stroke, tumor, multiple sclerosis, or dementia.

Finally, important data can be obtained from teachers, employers, and community agencies. Often, these sources have extensive, longitudinal data acquired over years of observation. Telephoning or personal interviews of these sources often reveal significant insights that would not have emerged otherwise.

## Types of Assessment Data

Comprehensive psychosocial assessment should yield a volume of data, which may be categorized along a variety of dimensions.

The analysis of premorbid history is extremely important. Many believe that predisability variables are extremely predictive of postmorbid adjustment. Educational background information consisting of level of education and specialized training, grades and test scores, reading ability, and interest in pursuing further education, may be helpful. For a child with a disability, information about past special educational programming or need for future special educational planning is very important. Information about family relationships should include data on the nuclear family unit and the history of the patient's capacity to relate effectively to various family members, including significant others like boyfriends or girlfriends. Evidence of marital or family discord or sexual dysfunction is also important.

Economic resources are of critical importance for the disabled person. Though most individuals have some type of health insurance, a disability imposes lifelong medical needs, such as intensive outpatient services and specialized equipment, which strain most people's finances. Assessment of the individual's eligibility for social security benefits is also crucial, since these benefits may become the chief source of income for many.

The home environment often poses great obstacles for disabled people: architectural barriers, access to transportation, and availability of medical services. Since most disabled people do not return to work for an extended time, review of avocational or recreational pursuits is a crucial task, though it is often neglected or downplayed in the early phases of rehabilitation. Social, religious, and community activities are also important for many individuals, because they provide a great deal of structure and emotional support.

Past behavioral patterns are often predictive of future coping behavior. Since many victims of trauma show a history of risk taking, sociopathic behavior, or evidence of alcohol or drug abuse, the examiner must ascertain whether these problems in fact preexisted, often by interviewing significant others. For the disabled child, a developmental history can yield useful information about the child's capacity to deal with stress. To help establish goals, an understanding of premorbid plans, hopes, and dreams is vital.

**Purpose of the Assessment**

Psychosocial assessment evaluates the specific nature of the disability, functional limitations, and prognosis. In the case of neurologic disability, the course of recovery may be difficult to predict precisely. However, this uncertainty and how the patient and family react to it can determine how realistic they are in future planning. The rehabilitation program may vary greatly if one is a 16-year-old with Duchenne muscular dystrophy, a 68-year-old with bilateral amputations, or a 21-year-old with cerebral palsy.

Since rehabilitation is heavily dependent upon learning and behavior, it is essential to ascertain how the disability has affected learning abilities and behavioral patterns. Information contained in the psychosocial assessment (especially when performed by a rehabilitation psychologist or clinical neuropsychologist) should reveal the patient's ability to attend, concentrate, learn, remember, process information, perceive the environment, follow instructions, respond emotionally, and act in a socially appropriate manner. If, for example, a stroke patient is assessed and observed to have an attention span of 5 minutes, to learn only through extensive demonstration and repetition, to follow only two-step commands, to display visual neglect, to tend toward emotional lability, and to behave in a very dependent manner, the evaluator has uncovered information that needs to be considered by the entire rehabilitation team in planning treatment.

Often, a patient's capacity to participate in and successfully complete rehabilitation is linked to his or her understanding of the disability and how it has affected self-concept, goals, and aspirations. For some individuals, the presence of a physical disability is a damaging blow to self-esteem and body image. Altered cosmesis and need to use adaptive devices is negatively viewed by society as being different, and such individuals tend to be readily rejected by others. Whether this is the case, individuals often feel that they are not whole persons or that they are not viewed as attractive. Feelings of decreased self-esteem are often part of a depressive reaction that is often seen in this population and is illustrated by the following case vignette.

*Case 1.* Patty, a 14-year-old girl, sustained a severe closed head injury in an automobile accident, resulting in diffuse brain damage. She sustained an initial loss of consciousness lasting 10 days, but managed to recover most of her physical abilities quickly. Within 3 weeks, she was able to leave the hospital and return home. She had some residual memory deficits, but, more importantly perhaps, her head was shaven, and she had scars on her hand and face. Prior to the injury, she had taken a great deal of pride in her appearance. These overt reminders of the accident were devastating. She refused to leave her house for 9 months after discharge from the hospital from fear of rejection by her peer group and feelings of low self-esteem.

Family systems and communication patterns are often disrupted by disability. For the families of neurologically impaired patients, the disruption and distress may be even greater than with other disabled persons, because their relative may "no longer appear to be the same person" due to the sometimes substantial alterations in cognitive and behavioral function. An area which is almost always affected is the preexisting roles of family members. In some cases, this involves alteration in the role of economic provider; in others, it means change in status roles: a formerly dominant, independent person may now appear passive and helpless. In their review of the research on social roles as they are affected by disability, Bishop and Epstein (1980) concluded that there may be a curvilinear relationship between level of disability and role conflict. With mild disability and uncertainty about outcome, role functioning may be somewhat disrupted. When disability is more severe and functional limitations are clear, the need to reallocate roles may be more evident so that less confusion is present. With profound, permanent disability, role demands on family members increase and family members often become overburdened.

A somewhat different, yet related issue to family adaptation is the role of the family in assisting with discharge planning, follow-up services, and placement, if necessary. Analysis of "family support" is critical and is usually obtained within the first few days after admission to a hospital setting, often by the social worker or case manager. Though many families appear quite supportive, their support is sorely tested by their willingness to assume responsibility for managing the affairs of their relative, providing a suitable living arrangement, or being involved in daily personal care or supervision.

Finally, one must assess the effects of disability on educational and vocational function. For many disabled people, the presence or onset of a disability has a significant impact upon their educational and vocational activities. For the child who becomes disabled, special accommodations may need to be made in his or her school program (aide, equipment, architectural modifications, special classes). Hopes of attending or completing college may be dashed when severe cognitive or behavioral problems are part of the disability. In many cases, individuals who engaged in heavy-duty blue-collar work must consider retraining and finding another vocation. One must assess the individual's interest in returning to these pursuits and integrate other assessment data into a determination of the viability of new educational or vocational goals.

**Assessment Methods**

Psychosocial data may be obtained in a variety of ways. The first step is to determine what the referral source wants to know. This may appear quite elementary but can be a persistent problem in rehabilitation, whether the referral comes from a physician, vocational rehabilitation counselor, or social service agency. Often, the referral will read, "Evaluate cognitive status" or "Assess family dynamics." This provides little clue as to the specific questions that need to be addressed. It is often necessary to return to the referral source and ask for a clearer definition of their questions. Examples of clearer referral questions are: "Given the patient's past behavior patterns, can the patient benefit from a comprehensive inpatient pain-management program?"; "How will a memory deficit affect the patient's ability to return to work?"; or "Please evaluate the severity of depression and the need for psychologic or pharmacologic intervention."

The next step usually involves reviewing all pertinent medical records. In the case of a physically disabled child, review of previous educational records is also quite helpful. If the patient is not currently hospitalized or has been referred from an external agency or private source, such information is often not readily available. Since past history is of utmost importance in establishing a premorbid baseline, the assessor should make every effort to obtain all of the pertinent data.

The clinical interview is the method of assessment most often used by psychosocial specialists. Though the specific focus or style of interviews may vary, whether conducted by psychologists, rehabilitation counselors, or social workers, the purposes remain the same—to establish good rapport, to gain an understanding of how disability has affected patients' views of themselves and their lives, and to determine the effectiveness of current coping methods. Often, a team member will interview members of the family or significant others; with a physically disabled child, this is particularly important. Usually, a developmental history can be obtained during the interview and an understanding of family interaction patterns (especially parent-child) may be acquired. The establishment of sound rapport within the initial interview is critical; it serves as the basis for a continuing relationship throughout the hospitalization and after discharge to the community.

Psychological, neuropsychological, educational, and vocational testing are frequently employed in the rehabilitation setting (see Lezak, 1983 and Athelstan, 1982 for extensive descriptions of these tests). It is important to recognize the limitations of psychological testing in a rehabilitation setting: (1) such tests sample a limited range of behavior in an artificial setting that may be only partly generalizable to real-world problems or situations and (2) many psychological tests are inappropriate for physically disabled persons because of sensorimotor or communication impairments and time restrictions. Despite these inherent limitations, analysis of test results can and often does provide useful information that assist in treatment planning. Perhaps the most widespread use of testing in rehabilitation is with neurologically impaired patients. Much information can be derived about neurobehavioral function that provides a profile of a patient's strengths and weaknesses. The psychologist is usually skilled at selecting specific tests that can yield the most useful information about patient function. Psychological testing is often held in high regard by physicians and others because it provides a quantitative index of psychological function, an area which is notoriously difficult to measure objectively. It should be stressed, however, that testing based on process yields qualitative observations about behavior that are of equal or greater value in understanding patterns of behavior. Unfortunately, consumers of psychological testing often ignore this fact and focus instead upon "the numbers."

Another method of analyzing and quantifying behavior is "behavior analysis" (Fordyce, 1982). In this method, a target behavior such as "poor motivation" is objectified into discrete, measurable units, such as failure to attend therapy sessions. Antecedent and consequent events are specified; a menu of possible reinforcers is identified; and a baseline, that is, a record of the frequency of the behavior under natural circumstances, is specified. When such data can be gathered, a behavioral modification program can be established to provide a treatment directly aimed at correcting the maladaptive behavior.

*Case 2.* Susan was a 42-year-old woman with a history of severe rheumatoid arthritis and chronic pain. She was admitted to the rehabilitation unit to improve her level of independent function. In the initial interview, it was noted that she displayed considerable "dependency behaviors" and derived great pleasure from reassurance and social reinforcement. Her "dependency" was broken down into a number of functional units: time out of bed, assistance needed for transfers, number of repetitions of strengthening exercises, initiation of behavior, and independent toileting. The frequency of these behaviors was recorded by the nursing and therapy staff over a 3-day period. With this data, a behavioral modification program was developed to increase the frequency of her target behaviors.

## The Written Report

A final step in the development of the psychosocial data base is the inserting of the assessment report into the medical record and providing it to the referral source. Too often, this report is filled with technical jargon, statistical data, abbreviations, or verbose language; it may be so lengthy and cumbersome that few professionals take the time to read and adequately comprehend it. Whenever possible, the examiner should be concise, eschew jargon, specifically address the referral questions, identify the nature and degree of problematic behavior, provide supporting data, and conclude with action plans that can be implemented in treatment.

## TYPICAL PSYCHOSOCIAL PROBLEMS

One of the most frequent psychological "disturbances" noted by staff in rehabilitation settings is depression. Depression may be defined as a diminished experience of pleasurable sensation manifested by feelings of loss, negativism, hopelessness about the future, social withdrawal, dull affect, crying, change in appetite, loss of interest in personal hygiene, and on occasion, suicidal ideation. Though most disabled individuals experience depression, it is rarely of the clinical variety and tends to diminish once a patient participates actively in a rehabilitation program. Active participation usually leads to increased independent function, thereby decreasing feelings of worthlessness, powerlessness, and pessimism. Suicidal ideation is not uncommon among newly disabled individuals, but suicidal gestures are relatively rare, except perhaps in the case of spinal cord-injured males.

The unmotivated or noncompliant patient is often viewed negatively by rehabilitation staff. Despite the best intentions, staff efforts to improve the physical and psychological well-being of the patient may be met with hostility or indifference. Examples of noncompliant behavior include not attending therapy sessions on time, refusing to work on certain therapeutic activities, not performing prescribed exercises on the ward, refusing to take medications, failing to maintain proper fluid or food intake, or avoiding making plans for discharge. Often the causes of poor motivation or noncompliant behavior are unclear. The patient is viewed as hostile, depressed, or "obnoxious" by staff members. It is important to try to ascertain the reasons for the behavior, which may be a conflict of goals, a failure to understand or acknowledge the disability, cognitive impairments, or conflicts with authority figures, such as staff.

The dependent patient appears content to assume the "sick" role and is perfectly willing to allow other caretakers, such as family or nursing staff, to perform activities he or she may be capable of performing alone. This behavior is frequently reinforced by family members, who view their disabled relative with sympathy and derive satisfaction in assisting with daily activities. Examples of dependency behavior include being fed by others, refusing to wheel the wheelchair, and failing to learn personal care routine. A classic example is seen in the overprotective parent who refuses to allow the child to do anything alone, despite reminders by staff that the child is quite capable of performing the activities without assistance. Sometimes, strict limits need to be placed on both child and parents if rehabilitation gains are to be realized.

The patient recently disabled due to illness or trauma is often quite anxious or tearful. this anxiety is manifested through excessive demands for attention, a need for continual reassurance, an overly cautious attitude, a tendency to overreact to pain and other physical symptoms, and a resistance to attempt new tasks for fear of failure. These reactions are quite common but often shortlived. For patients with a premorbid history of anxiety, however, the disability can evoke tremendous apprehension, feelings of uncertainty abut the future, and continual worry.

*Case 3.* Bill was a 62-year-old male who was found lying in his apartment after apparently suffering a stroke. When medical stability was achieved, he was transferred to the rehabilitation unit. On the second day after admission, the therapists reported that the patient was extremely anxious and virtually refused to engage in any task presented to him in physical therapy. In an interview with the rehabilitation psychologist, the patient displayed a very high level of anxiety, acknowledging that he was very anxious prior to the stroke but that his anxiety was now greater than ever. He was scared of having another stroke and becoming helpless. He was fearful of physical therapy because he felt as if physical activity would induce another stroke.

Unrealistic thinking or denial is often observed in cases of neurological illness wherein the course of recovery of function is quite uncertain. Sometimes the recognition of the permanence of disability is too painful for the patient to contemplate. To avoid dealing with this reality, a patient may refuse to accept that certain physical or mental limitations exist. Though the patient may cooperate with the rehabilitation program, he or she may not accept recommendations for outpatient treatment, may fail to take prescribed medications, or may avoid making appropriate discharge plans (e.g., returning to a second-floor apartment despite the fact that a wheelchair is required). Another common manifestation of denial is the refusal to relinquish premorbid aspirations. For example, a 24-year-old college graduate who was intent on attending law school suffered a head injury resulting in severe cognitive impairments. Despite recommendations to the contrary, he proceeded to apply to law school and committed suicide when he was rejected.

Finally, there are a great variety of cognitive and behavioral impairments resulting directly from brain damage. Inability to learn, dysfunctional memory, disinhibited behavior, impaired language and communication, inability to read or write, lack of initiative, and uncontrolled aggression are a few of the symptom patterns. Specific interventions such as cognitive remediation, behavioral management, or pharmacologic intervention may be necessary. Because the patient's personality has changed so much, it is extremely important to educate the family as to the nature of the neurologic insult, the prognosis, and the ways they can be most helpful in dealing with problem behaviors.

## SUMMARY

This chapter has described the process of psychosocial assessment and its role in the total rehabilitation of the physically disabled child and adult. Psychosocial assessment is an essential component of the initial interdisciplinary evaluation of the disabled person. Successful rehabilitation must consider not only the nature of the specific physiologic impairment but the relationship of the impairment to the disabled person, the family, and the social environment.

Adaptation to disability varies from individual to individual. Past research has failed to demonstrate a consistent relationship between personality type and disability. Reactions to disability, whether in the form of depression, inertia, anxiety, fearfulness, or the like are not time limited. Disability may impose both acute and chronic stress upon the individual. Family members can often greatly influence the rehabilitation process.

Psychosocial assessment is usually conducted by designated members of the rehabilitation team, such as psychologists, social workers, or rehabilitation counselors. However, it is not the sole province of these professionals and should be considered a part of the initial evaluation by other members of the rehabilitation team as well. Psychosocial assessment uses multiple sources of data: past medical, vocational, and educational records; premorbid patterns of behavior; learning ability; nature of the disability; and acceptance of the disability and its limitations. Specific methods of assessment include the interview with patient and family; psychological, neuropsychological, and vocational testing; and systematic behavioral analysis.

The adequacy of current rehabilitation methods in restoring the disabled individual to a productive position in society cannot be based solely on objective measures of functional improvement in mobility, strength, or in activities of daily loving. In addition, these efforts must be judged on the capacity of the individual to effectively cope with the stresses associated with having a disability and achieving a satisfactory quality of life. Effective psychosocial assessment and timely interventions are necessary elements in the rehabilitation process if quality of life is to be the ultimate measure of successful rehabilitation.

## REFERENCES

Athelstan, G. (1982). Vocational assessment and management. In F.J. Kottke, G.K. Stillwell & J.F. Lehmann (Eds.), *Krusen's handbook of physical medi-*

cine and rehabilitation. Philadelphia: W.B. Saunders, 163–189.

Bishop, D.S. & Epstein, N.B. (1980). Family problems and disability. In D.S. Bishop (Ed.), *Behavioral problems and the disabled*. Baltimore: Williams & Wilkins, 337–364.

DeJong, G. (1979). Independent living: From social movement to analytic paradigm. *Archives of Physical Medicine and Rehabilitation, 60,* 435–446.

English, W. (1983). *The role of the family in rehabilitation*. Rehabilitation Research Review. Washington, D.C.: National Rehabilitation Information Center.

Fordyce, W. (1982). Psychological assessment and management. In F.J. Kottke, G.K. Stillwell, J.F. Lehmann (Eds.), *Krusen's handbook of physical medicine and rehabilitation*. Philadelphia: W.B. Saunders, 151–162.

Greif, E. & Matarazzo, R.G. (1982). *Behavioral approaches to rehabilitation*. New York: Springer.

Guidubaldi, J., Kehle, T.J. & Murray, J.N. (1979). Assessment strategies for the handicapped. *Personnel and Guidance Journal,* 245–251.

Lezak, M. (1983). *Neuropsychological assessment* (2d ed.). London: Oxford University Press.

Pelletier, K.R. (1977). *Mind as healer, mind as slayer: A holistic approach to preventing stress disorders*. New York: Dell/Delta Books.

Sakinofsky, I. (1980). Depression and suicide in the disabled. In D.S. Bishop (Ed.), *Behavioral problems and the disabled*. Baltimore: Williams & Wilkins, 17–51.

Selye, H. (1974). *Stress without distress*. New York: New American Library.

Shontz, F. (1977). Six principles relating disability and psychological adjustment. *Rehabilitation Psychology, 24,* 207–210.

Trieschmann, R. (1980). *Spinal cord injuries: Psychological, social and vocational adjustment*. New York: Pergamon.

Versluys, H.P. (1980). Physical rehabilitation and family dynamics. *Rehabilitation Literature, 41,* 58–65.

# 4

# The Problem for Neuropsychologists: Differentiating Physical, Emotional, and Cognitive Abilities

*Ralph J. Kiernan*

## HISTORICAL BACKGROUND

Throughout the first half of the twentieth century the psychology of brain function was heavily influenced by the elegant experiments and antilocalizationist theories of Carl Lashley (1929). He showed that complex behaviors are impaired by cerebral damage but that such effects are dependent on the degree rather than the location of damage. The brain was seen as equipotential, with all but specific sensory and motor areas contributing equally to the performance of complex tasks.

It is only natural that a unitary concept of brain dysfunction, *organicity*, should emerge from this source. The concept of organicity is based on the "assumption that any and all kinds of brain damage lead to similar behavioral effects" (Davison, 1974). The effects are vaguely stated and include concrete thinking, reduced level of performance, and difficulty with certain types of intelligence tests. This unitary view was powerful and led to certain conceptual difficulties that continue to muddle our thinking about brain dysfunction to the present day (Davison, 1974). For my purposes, it is important only to trace the influence of the concept of organicity on neuropsychology.

In the 1930s and 1940s there was the hope that a single sensitive test would be able to identify the unitary deficit syndrome, organicity. The Bender/Gestalt Test (Bender, 1938) and the Porteus Mazes (Porteus, 1959) were two such tests that were believed to have sensitivity to brain dysfunction. Indeed, for many years the Bender/Gestalt Test was taught as the one instrument to add to a standard psychological assessment battery when organicity was suspected. It was only natural for early neuropsychologists to select tests with sensitivity to brain dysfunction regardless of the extent or location of dysfunction.

Halstead (1947) was primarily interested in frontal lobe functioning, and when he selected 10 tests from his original 27, he based his selection on the ability of these tests to discriminate frontal lobotomy subjects from control subjects. Ralph Reitan, however, was far more concerned with the general task of identifying brain damage. His validity studies of the evolving Halstead-Reitan Battery (Reitan, 1955a, 1955b) clearly established the diagnostic utility of neuropsychological testing, but they did little to alter the unitary notion of organicity. Indeed, the Impairment index (Halstead, 1947) is based on the assumption that all tests in the battery can be summed to make an overall prediction about probability and severity of brain dysfunction. While there is little question that the Halstead-Reitan Battery can be used to localize brain lesions (Reitan, 1957), the initial tests in the Halstead Battery were all selected and validated in terms of overall sensitivity to brain dysfunction.

It is important to have highly sensitive tests in any neuropsychological test battery. Such tests signal us when there is any possibility of underlying brain dysfunction; they heighten our concern and cause us to intensify our investigation when subjects perform poorly on them. Sensitivity, however, is a two-edged sword.

If we were to construct a test battery solely of tasks which had proven themselves to be highly sensitive to brain dysfunction regardless of location, we would have tasks that all measure the same thing. They would be complex, multifactored tasks that demand the coordination of many skills for correct, efficient solution. Halstead's Tactual Performance Test (Halstead, 1947) is an excellent example of such a task. This test requires the placement, while blindfolded, of 10 geometrically shaped blocks into a form board. The subject is asked to complete the form-board task, first by using the dominant hand alone, then the nondominant hand alone, and then by using both hands. The score is based on the total time for completion of all three trials. The specific skills involved include planning and organizing functions, spatial abilities, memory, and both left and right hand sensorimotor functions. This task is challenging for even good performers, and it is not surprising that it has shown itself to be one of the most sensitive tasks in the Halstead-Reitan battery (Reitan, 1964). It is possible of course to derive a great deal of useful observational data from this task and to make intertrial comparisons (Lezak, 1983) that have great significance. Yet, if one had a test battery consisting of only sensitive tests, one would have to rely on qualitative observation in making either localization inferences about brain damage or specific statements about the nature of the disability. One cannot, however, expect to reliably derive much specific information from observations. Even the most basic question of presence versus absence of brain dysfunction may be difficult to answer with sensitive tests alone.

In the Wechsler Adult Intelligence Scale, Digit Symbol (Hirschenfang, 1960) is the most sensitive subtest to brain dysfunction. This is probably less attributable to its great complexity than to the simple fact that it is the only subtest of the Wechsler scale in which speed of performance directly determines score. It is, however, sensitive to the psychomotor retardation seen in depressed patients (Rapaport, 1945) and to normal aging (Savage et al., 1973). Thus, a highly sensitive test may be too sensitive and be affected by many factors not related to brain dysfunction. Such tests quickly lose their efficacy when depression and brain dysfunction coexist. Our qualitative observations may simply confirm that the performance was slow, without aiding us in diagnostic inference.

There is an important place in neuropsychology for tests that are not highly sensitive to brain dysfunction. Such tests fail to discriminate overall populations of brain-damaged subjects from normal subjects, because they are designed to assess only a very specific skill. For example, the Token Test (De Renzi & Vignolo, 1962) is a measure of oral language comprehension, which is usually performed well by both normal subjects and patients with mild brain dysfunction. Yet it is exquisitely sensitive to language comprehension problems when they are present (Boller & Vignolo, 1966) Its false-positive rate (i.e., the percentage of patients scoring in the impaired range who do not have brain dysfunction) is extremely low (Benton & Hamsher, 1978). Its true-positive rate is also low except in aphasic patients. Tests such as this do not produce a high yield, but they do provide specific information when positive. The neuropsychologist must balance the power of highly sensitive tests with the selectivity of highly specific tests.

Nowhere is this balance needed more than in evaluating physically disabled patients who have secondary emotional problems. Often, direct physical limitations will prohibit the use of certain multifactored, highly sensitive tests altogether. For example, weakness and difficulty with coordination make smooth performance and meaningful timing of the Tactual Performance Test impossible. There is the ever-present danger that a mild sensory or motor deficit will intrude on complex test performances in a manner that was not anticipated. Such subtle intrusions may not be appreciated by the examiner at all. For example, minimal right-visual-field neglect will result in slower times on the Trail Making Test (Armitage, 1946; Reitan, 1955b), and such slow times may easily be misinterpreted as problems in sequencing. The challenge for neuropsychologists in evaluating disabled patients is largely in identifying those tests that allow the patient the most unimpaired avenue for performance. The challenge becomes much greater when these patients also have disabling emotional factors (e.g., anxiety and depression). Anxiety and depression are common in physically disabled patients, and they often exert subtle effects on test performances, which are easy to misinterpret. At times, marked anxiety and/or depression are present, and all tests are profoundly affected. It is not possible to do routine, standardized neuropsychological testing in such cases. For example, many patients with minor head trauma develop severe psychophysiologic symptoms marked by anxiety, depression, and an all-consuming concern and preoccupation with somatic symptoms. With such patients, the clinician must be ever vigilant for interfering effects of emotional factors on neuropsychological tests. Great care must be taken to tease out the emotional factors in testing by finely controlling and closely monitoring the attention and concentration demands of various tests. The rote administration of a test battery and routine interpretation of scores doom the clinician to erroneous diagnostic inferences.

The challenge for the neuropsychologist working with patients with physical disabilities and secondary emotional problems is to put aside the fixed test battery with its automatic, cookbooklike interpretations. Only then can the complex detective job of unraveling the tangle of factors that influence test performances in these patients begin.

## PHYSICAL LIMITATIONS

Patients with physical disabilities are often unable to perform tasks that require a motor response. If they are able to perform the motor task, albeit slowly, how will we interpret the results?

Many neuropsychological tests derive their sensitivity from the fact that they are timed. Patients with specific cognitive processing problems are unable to determine and use the correct strategy for solving the task quickly and efficiently. They may well solve the task eventually, but their performances are much slower than average. Thus, a test score based on time directly reflects the degree of difficulty. This is not true in the patient slowed by physical disability. If we simply use the score alone, we will fail to distinguish the two causes of slowed performance. We will misinterpret the low score as reflecting cognitive disability rather than physical disability.

We could elect to omit such timed tasks altogether from our test battery, but this would overly restrict us and eliminate too many useful and informative tasks. For example, when testing a patient with mild weakness and incoordination associated with multiple sclerosis, we may want to use the Trail Making Test. Yet the times will surely be slow because of physical disability. Fortunately, the Trail Making Test has a built-in control, which helps us escape this dilemma. Both parts A and B of this test require identical motor-speed ability. Part B, however, requires far more complex sequencing than A. Because of this, part A can be used as a motor-speed control for part B, thus allowing a more direct observation of the sequencing factor itself. There are no norms for doing this. We have taken a well-standardized neuropsychological instrument and used it, because of known physical disability, for qualitative observation only. We may have difficulty using the information, which we obtain in such a nonstandard fashion, but we will at least be mindful of the diagnostic problem that we are facing. In these cases, the naive use of the Trail Making Test with its usual standardization would obviously be disastrous.

Most neuropsychological tests do not have built-in controls like the Trail Making Test's. A test such as the Digit Symbol subtest of the Wechsler Adult Intelligence Scale (Wechsler, 1955) might well be impossible to use meaningfully with multiple sclerosis patients. We need to clarify which cognitive factors we are interested in, and we must select tests that separate these factors from motor speed. For example, the Raven Progressive Matrices (Raven, 1960) is a powerful measure of reasoning ability, which is largely nonverbal in nature and is completely independent of motor speed. A recent patient of mine had received traumatic damage to his left hemisphere, cerebellum, and right eye. As a result he was globally aphasic, had spastic use of only his left hand, had a complete right homonymous hemianopsia, was blind in the right eye, and was extremely slow to respond. Yet, he was able to slowly go through the Raven matrices and obtain a 50th percentile ranking. This test alone demonstrated a surprising degree of reasoning ability that had been unsuspected.

The primary rule in assessing the patient with motoric slowing secondary to physical disability is to stay mindful of the potential effects of the disability on each test used. At times we will simply need to avoid some of the most sensitive and trusted tests because of this unavoidable factor. At other times, we may choose to use the test but in a qualitative fashion, avoiding standardization norms, which are likely to be misleading. Lastly, we can select, from the large number of test instruments available, those measures that directly assess cognitive factors while avoiding the motor-speed factor.

The approach to testing the slow patient can be used as a model for assessing all patients with sensory and/or motor disabilities. This model includes three options: (1) to avoid tests that are excessively affected by the sensory or motor disability, (2) to use tests that involve the disability in a nonstandard, qualitative fashion, and (3) to find tests that circumvent the physical limitation altogether. These three options will be discussed below for each of the specific sensory limitations.

**Visual Limitations**

There are many ways of testing patients who have visual limitations related to the eye itself. Visual acuity can be measured and test material adjusted accordingly. Many tests can be presented in relatively standardized ways to the patient with mild acuity problems by simply using large-print materials. For example, the University of Wisconsin's Neuropsychology Laboratory has used a large-print version of the Wide Range Achievement Test's reading and arithmetic sections for many years. The visual demands of nonreading tests are often harder to judge. However, most patients with sufficient acuity for reading can perform most visual tasks. Timed or highly speeded visual tasks (e.g., the Spatial Relations subtest of the Primary Mental Abilities Test [Thurstone, 1938]) will be affected in subtle and unspecifiable ways that may prohibit their use in these patients. Others, such as the Trail Making Test, may be used with more subjective than standardized interpretation. Tests that require fairly intricate analysis (e.g., the Picture Arrangement subtest of the Wechsler Adult Intelligence Scale (Wechsler, 1955) may be attempted but with close monitoring of the difficulties the patient reports. For example, item 8 of the revised test (the king-fishing story) (Wechsler, 1981) is difficult to arrange correctly unless the diver at the end is correctly recognized.

In the blind patient, the examiner's task is actually simpler. Here it is obvious that all visual materials must be avoided, and the examiner can proceed to assess cognitive skills in other ways. There are certainly limitations to such an assessment, and many favorite tests, for example, the Wechsler

Block Design subtest (Wechsler, 1955) and the Benton Visual Retention Test (Benton, 1974) cannot be used. In addition, certain areas of skill are difficult to evaluate altogether, for example, spatial memory and nonverbal reasoning.

Visual problems associated with cerebral lesions are more difficult to take into consideration when testing, yet they represent an area of disability that must be directly assessed in its own right. Scotoma or blind spots in the visual field due to a circumscribed occipital lobe lesion are usually mapped by the neurologist using confrontation testing or more precisely with a tangent screen. Scotoma usually do not interfere with neuropsychological testing unless they become quite large or they are associated with hemifield neglect.

The patient with a homonymous hemianopsia has no vision for the visual half-field on the side opposite the brain lesion. Such patients have normal vision within their intact half-field but are completely unresponsive to visual stimuli in their blind field. Similarly, half-field neglect can be observed in patients with posterior brain lesions that do not produce scotoma. These patients may fail a visual task simply because a portion of the stimulus material lies within the neglected half-field. For example, the pictures from the Picture Arrangement subtest of the Wechsler Adult Intelligence Scale are arranged in a left-to-right array, which can in itself be quite challenging to a patient with neglect. It is not unusual for a patient to fail an item completely, because of not seeing all of the pictures involved, or to at least be slowed down considerably by the task of visually locating all of the pictures. A similar array of picture choices appears on the Raven Progressive Matrices Test. Tests such as the Trail Making Test become largely measures of visual search ability in such patients as the sequencing demands of the task become overshadowed by the visual neglect. A visual memory task such as the Benton Visual Retention Test has three figures arranged from left to right so that patients with neglect often fail to perceive one or the other end design. These tests can be used directly to assess the degree of neglect, but the interpretation of other skills (e.g., nonverbal reasoning, sequencing ability, and memory) is often not possible.

**Auditory Limitations**

The effects of auditory limitations on neuropsychological assessment are discussed separately for adults and children. There is usually little problem in evaluating the hard-of-hearing adult. Often the problems encountered are overcome simply by speaking more slowly and distinctly. When hearing problems are more severe, it is possible to present test instructions and test material in written form. There are very few tests that cannot be satisfactorily administered in this manner. Using a sign language interpreter, especially for prelingually deafened persons, is often a more optimal solution. Certain tests that depend on auditory presentation, such as the Seashore Rhythm Test (Seashore et al., 1960) or the Halstead Speech Sound Perception Test (Halstead, 1947), may not be possible, but their omission represents only a small loss in testing power.

In children, however, severe auditory problems may have a profound effect on testing. Language development is clearly delayed in this population, and standard language tests can be used to rate the degree of delay. But these tests do not help decide whether the delay is due only to deafness or whether some of it is due to central language-processing problems. This complex assessment problem, however, is beyond the scope of the present chapter. (Elliott et al., 1987; Heller & Harris, 1987; and Zieziula, 1982 provide more detailed information.) Some nonlanguage tests that superficially appear to be appropriate for deaf children lack normative data for the deaf and may give misleading scores when the normal standardization is used. Others, however, have been normed on samples of deaf children, such as the Wechsler Intelligence Scale for children–Revised (Wechsler, 1974), normed by Anderson & Sisco (1977). The Leiter International Performance Scale (Leiter, 1969) was specifically designed to circumvent this difficulty by assessing a variety of nonverbal skills without any use of language and by providing norms based on a population of deaf children. This test frequently gives average or above average scores in children who are below average on the nonverbal portions of the Wechsler scale. Each standardization group (normal vs. deaf children) has its own limitations and must be used carefully in drawing inferences.

## EMOTIONAL AND MOTIVATIONAL LIMITATIONS

The physically disabled frequently manifest secondary emotional and/or motivational problems, which pose the most difficult challenge to adequate neuropsychological testing. These problems are especially challenging because they can affect performance on any test, and the effects may not be obvious during the testing session. When such factors are combined with sensorimotor limitations, the already difficult assessment task is further complicated. These psychological limitations, however, are present to some degree in a broad spectrum of patients referred for neuropsychological testing. In many cases, the symptoms that potentially represent a neurologic lesion (e.g., impaired memory) are solely the result of psychological factors. The challenge to neuropsychologists in assessing such cases is a fundamental problem for neuropsychology. The discussion to follow therefore has application well beyond the present focus on patients with physical disabilities.

## Attention

William James (James, 1958) stated that "everybody knows what attention is," but, in fact, the term proves extremely difficult to define. Webster's defines *attention* as "applying the mind to an object of sense or thought." "Applying the mind," however, is an internal process that cannot be observed directly. We use a variety of indirect behavioral observations in judging a person to be attentive. But what is attention itself? Attention is an active process that can only be judged by the act of engaging in some task. It is obviously fundamental to test taking, and it must always be considered when test performances are interpreted. Once attention is no longer considered as a conceptual abstraction but as an act, it becomes clear that many factors can disrupt or interfere with this act. I consider attention the "final common pathway" for all indirect effects on testing. By this I mean that attention is the medium through which such diverse factors as anxiety, depression, mild delirium, and thought disorder affect test performances. These factors all interfere in their own ways with the active process of attending to a task.

A baseline level of attending capacity is necessary for meaningful testing to even take place. The patient must look at or listen to the stimulus materials presented and make at least a minimal attempt to carry out the task requirements. Without this much attention, testing is impossible. Often, however, the patient will give this much attention but no more. In these cases tests can be administered, but the performances and scores defy meaningful interpretation.

I have had occasion in the past few years to review a number of detailed neuropsychological assessments of patients with profound posttraumatic psychological syndromes. Each of these cases was characterized by uniformly poor test performances, which lacked the highly variable pattern of abilities and disabilities typical of traumatic brain damage. The unusually low test scores, usually associated with minor trauma, were in fact unbelievable. In these cases the previous neuropsychologists described the significant emotional effects that the traumatic events and their associated physical symptoms had had on patients. They even discussed the breakdown in attention and concentration that was observed at various points during the testing sessions. They did not, however, realize that their hours of neuropsychological testing produced results that were uninterpretable, except as documentation of the patients' inability to sustain adequate attention.

Such cases are not rare in neuropsychology, and the difficulty that they present to the examiner is apparent in the many confusing interpretations rendered. A fundamental rule for neuropsychology is, "Bad scores alone do not imply bad brain." Indeed, in many of the cases I reviewed, the neuropsychologist simply took each score, one at a time, and made a direct interpretation about a certain area of the brain. It matters little in such cases which test the clinician used since the problem lies not in the tests and scoring criteria themselves.

Sometimes it is most helpful to focus on a simple task and to ask simple questions about performance. One such task is the Knox Cube Test, which is included in the Arthur point scale of nonverbal performance tests (Arthur, 1947). The Knox Cube Test consists of four 1–1½-inch cubes arranged in a row on a flat base, with approximately 1–1½ inches between each block. The tester taps a certain number of blocks in a preset order, and the patient must repeat the exact tapping sequence. The task begins with a simple two-block tapping sequence and increases gradually to a maximum of seven-block taps. This is a test of visual attention span and little else. It is not sensitive to brain dysfunction, and only rarely is a patient seriously impaired on this task. When it is poorly performed, this invariably has special meaning.

A patient of mine began her testing session by failing miserably on the Knox Cube Test. She was a middle-aged woman who was having increasing problems with memory and had been concerned about her ability to continue working in an administrative job. She was clearly worried about herself and mildly depressed with her present situation in life. She had great trouble on the Knox Cube Test, missing several sequences at the relatively simple, 3- and 4-block tapping level. The balance of her test results was mixed, with several fine performances on demanding tests of reasoning ability and only a few other instances of significant attentional failure. She had difficulty on memory tasks that required highly focused attention to a single presentation of stimulus material (e.g., two story paragraphs from the Wechsler Memory Scale [Wechsler, 1945]), but she showed excellent memory on a demanding 20-word-list learning task administered with the selective reminding technique (Buschke & Fuld, 1974). There was no consistent evidence of specific disability, and the striking instances of attentional failure seemed directly related to her anxiety and concern about herself. The dramatic, poor performance on the Knox Cube Test served to underscore the severity of her attentional blocking and to make understandable her perceived job-performance difficulties.

A second, slightly older woman also came to testing because of increasing problems with memory. Since she performed no work outside of the home, it was difficult to get an accurate, independent estimate of how much difficulty she was having. She did, however, dramatically fail the Knox Cube Test and became confused at the simplest levels. The balance of testing, however, was quite different from the first case. This woman had difficulty with many of the tests administered. Her scores were not uniform in that she had much more difficulty on the more challenging tests. She clearly had generalized, severe cognitive impairment. She became easily confused by even slightly complex tasks, and she was unable to demonstrate any useful memory. In her case, the Knox Cube Test underscored the severity of her generalized intellectual decline. She failed not because of specific attentional blocking but because of the reduced attention span associated with her severe dementia syndrome.

These two cases illustrate the range of cognitive conditions that can underly total failure of a very simple attention-span task. It would be a serious error to interpret the scores as reflecting a visual attention-span deficit in each case and then ascribe such a deficit to dysfunction in a specific area of the brain such as the posterior right hemisphere. In each of these cases, one must first ask the basic question of why this patient had such marked problems on a very simple task. In the first case, the poor performance was not associated with a specific deficit in visual attention span but an intermittently demonstrated attentional disturbance. This was clearly secondary to anxiety blocking, and this patient was able to demonstrate excellent cognition and memory as well as fairly good attention at other times. Nor did the poor performance represent a specific visual attention deficit in the second patient. Her cognitive deterioration had progressed to such a degree that even relatively simple tasks became complex and confusing to her. These two patients, despite their poor Knox Cube Test performances, were readily able to complete a number of other tests that were diagnostically useful.

There is another type of patient who might fail a simple task such as the

Knox Cube Test. In this patient, the interfering effects of a very severe depression or the disorganizing effects of a psychiatric thought disorder are so marked as to undermine even minimal attention and sustained effort. The test scores obtained in such cases cannot be interpreted directly without the strong possibility of a false-positive interpretation of brain dysfunction. These patients are untestable! This was exactly the case with the posttraumatic psychological syndromes I described previously. Without a minimal degree of attention and cooperation from the patient, neuropsychological assessment is impossible. The careful neuropsychologist must be willing to draw the conclusion that (1) testing is not possible for the present because of the interfering aspects of the patient's depression or thought disorder and (2) it might be useful to attempt reassessment in the future, when there has been improvement in the patient's clinical condition.

If simple tests such as the Knox Cube Test can be highly affected by fairly mild but fluctuating anxiety, it should be obvious that complex and demanding neuropsychological tests are even more affected. How can we expect a severely depressed patient to perform the demanding stereognostic problem solving required by Halstead's Tactual Performance Test (Halstead, 1947)? Such highly complex tasks are frustrating for the average performer, and they are doubly so for the severely depressed. In addition, this test bases its score on time required for completion, and it derives much of its sensitivity from this. Timing, however, becomes meaningless when there is significant psychomotor retardation. I tested one very depressed man who performed the Tactual Performance Test in what appeared to be slow motion, with many pauses and deep sighs punctuating his effort. He was eventually able to complete the three trials of the task, but his times were very slow and extended well into the impaired range. No standardization sample will help us interpret such a performance. Yet the very factors — complexity, speed, flexibility in thinking — which make a task sensitive to brain dysfunction also make it sensitive to conditions that interfere with sustained attention. If neuropsychologists are to develop clinical skill we then must be constantly struggling to distinguish indirect effects mediated by attention from the direct effects of cerebral dysfunction. At present, the most common and glaring error in neuropsychology is misinterpreting the indirect effects of attention as reflecting brain damage.

**Assessment Strategies**

Since attentional problems are commonplace and often present in patients who are referred for neuropsychological assessment, it is essential to have an array of strategies for dealing with them. Indeed, the wisdom of the neuropsychologist will reveal itself in the cleverness with which such interfering factors are handled in the course of testing. I would like to outline and discuss four strategies:

1. Assess the attentional factors directly.
2. Eliminate attentional factors as much as possible from the cognitive variables of interest.
3. Compare and contrast each test with related tests to obtain additional diagnostic information about the respective roles of cognition and attention.
4. Restructure test instructions and test items to obtain more insight into the nature of the failed performance.

It is not possible to test attention in isolation. One must always attend to something, and there must always be some specific task demands. The Knox Cube Test described earlier is, after all, a visual task that involves following a moving stimulus across the midline and then carrying out the act of repeating the tapping sequence. It will obviously be affected by severe visual acuity problems, but more subtly it will be affected by visual half-field blindness and/or neglect. Substantial motor incoordination, tremor, or dysmetria will also hamper performance of this test. Nevertheless, it involves few, if any, problem-solving skills, and it makes no demands on memory beyond simple attention span. When it is not compromised by visual or visual field problems, it is a relatively pure measure of sustained attention. The Digit Span subtests of either the Wechsler Adult Intelligence Scale (Wechsler, 1955) or the Wechsler Adult Intelligence Scale–Revised (Wechsler, 1981) similarly measure attention. Admittedly, they may be affected by language-processing problems in either verbal comprehension or repetition, and they depend on a certain level of auditory acuity and speech. They otherwise demand little more than attention. Although other tests of attention have been devised (Bender, 1979; Diller et al., 1974; Gronwall & Sampson, 1974; Smith, 1968), the Knox Cube Test and the Digit Span Test have the advantages of being simple and easily administered and of having a broad range of adequate performance in the normal population.

Assessing attention directly is as impossible as eliminating the need for it completely: a patient must bring minimal attention to any task in order to take and complete it. Some tests, however, are highly demanding of specific cognitive skills and minimally demanding of attention. Untimed reasoning tests, for example, the Raven Progressive Matrices (Raven, 1960) or the written form of the Gorham Proverbs Test (Gorham, 1956) reduce much of the attentional demand by removing all time constraints. On these tests a patient's intermittent attending and occasional distractions do not prevent slow but steady progress. In memory testing, repeated learning trials like those included in the selective reminding procedure (Buschke & Fuld, 1974) or the Rey Auditory-Verbal Learning Test (Rey, 1964) permit the inattentive patient to learn the words eventually, despite attentional shortcomings.

Although one can neither assess attention in pure fashion or eliminate it entirely, one can identify the relative contributions of specific cognitive skills and attention to any given task. In the area of verbal memory, the Digit Span Test (Wechsler, 1981) can be represented in one corner (high attention, low skill) of a two-dimensional space plotting degree of attention against degree of new verbal learning ability (figure 4.1). The two tests mentioned above, the selective reminding procedure and the Rey Auditory-Verbal Learning Test are located in the opposite corner (low attention, high skill). Somewhat more affected by attentional factors but still depending heavily on learning ability is the Paired Associate Learning subtest of the Wechsler Memory Scale (Wechsler, 1945). This test requires attention, especially to the difficult or "hard" pairs, but its three repeated trials help to overcome mild attentional deficits. Its scoring system, by weighing equally all three trials, unfortunately blends the attention and learning aspects of this task. The Logical Memory subtest of the Wechsler Memory Scale is much more of a primary attention task in that only one presentation is given of a fairly complex story paragraph. The patient who is prone to attentional blocking is often overwhelmed by the amount of material presented. Indeed, the majority of patients that I have tested who scored zero on either paragraph have not had memory disorders but rather attentional blocking. It is not always so simple, however, to distinguish the relative role of cognitive factors and attention.

This brings us to point 3, which involves comparing the results of one test with another. Such intertest comparisons are essential to all reasoning in neuropsychology. It is always necessary to test hypotheses based on performance on any single test with predicted performances on related tests. It would be reassuring to know that a patient who obtained a very poor score of 3 on the Logical Memory Paragraphs of the Wechsler memory Scale had gone on to master a 20-word selective reminding list adequately in five to six learning trials (see figure 4.1). In the absence of specific language disability, it would reassure us that the poor performance on the story paragraphs was more likely due to attentional blocking than to problems with learning and retention. If this same patient showed attentional problems on the early trials of the selective reminding task, it would lend even further

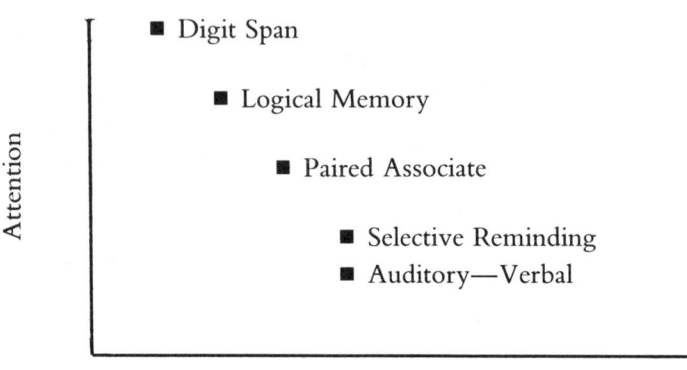

**FIG. 4.1.**
Schematic Plot of Attention vs. Verbal Learning for 5 Memory Tests

support to such a hypothesis. The following case shows how these two tests can be used in relationship to each other and to the Digit Span subtest of the Wechsler Adult Intelligence Scale.

A 27-year-old man had a renal transplant 3 years prior to testing. Although his renal functions were stable at the time of testing, he had been experiencing memory problems in his college courses. He was of average intelligence (full scale WAIS I.Q. = 96), but he had not done well in high school. He appeared very anxious on all tests that required sustained attention, and he consequently had difficulty with the Digit Span subtest (digits forward = 5, digits backward = 4) and the Logical Memory subtest of the Wechsler Memory Scale (average story recall = 5). He had no difficulty in mastering a 20-word list given with the selective reminding technique. After six trials he recalled 16 of the 20 words, and he consistently recalled 11 of the 20 words on each of the last three trials. These are excellent scores showing good learning ability. Additional testing failed to show any evidence of a specific language disability, and his subjective complaint of memory dysfunction was attributed to anxiety, which probably resulted from the impossible challenge that college-level course work presented to him.

Neuropsychologists cannot be content to take the role of the psychometrician, simply collecting a set of test scores. They must be actively involved in the testing process and dedicated to ferreting out the underlying cause of a poor test performance. This is an inherent weakness of the highly standardized western approach to testing and assessment. Luria's method (Luria & Majovski, 1977; Spiers, 1981) as distinguished from the so-called Luria-Nebraska Neuropsychological Battery (Golden et al., 1980) involves the active exploration of test performances, with constant formation and testing of hypotheses. The testing process is a dynamic one in which tests are selected, used, and followed up on, depending on the results obtained. Spiers (1981) points out that while Luria was fond of arithmetic problems as a diagnostic tool, he did not simply score each problem as right or wrong and accumulate total or average scores. Instead, each arithmetic problem represented a starting point from which one could explore the patient's attentional, linguistic, reasoning, and calculating abilities. By varying a single problem and extending the limits, a whole set of hypotheses could be formed and tested. This is the attitude I advocate when I suggest restructuring tasks and testing limits.

Let us take as an example the Arithmetic subtest of the Wechsler Adult Intelligence Scale (Wechsler, 1955). This test can be easily affected by attention (Lezak, 1983), and it is part of a triad of Wechsler subtests (Digit Span, Digit Symbol, and Arithmetic) that represent an attentional or "freedom from distractibility" factor (Berger et al., 1964). The test consists of a series of word problems that require the person taking the test to understand all of the information provided, to reason out the correct strategy, and then to carry out a one- or two-step computation. From a neuropsychological point of view, there are many places where the patient's test performance can break down. The patient may fail to attend to the question as it is read, to comprehend the complex verbal problem, to sort out the correct strategy, to correctly perform the computation, or to keep track of the problem posed while performing the calculation.

Since my primary concern in this section is with attention, I analyze here only one aspect of the task. It is fairly common for a patient to ask to have a problem repeated, and this is allowed by the Wechsler manual (Wechsler, 1955) providing that the response is given within the generous 60- or 120-second time limit. If we record the number of times the patient asks to have a problem repeated, we will be further ahead in reasoning than if we simply depend on the overall score. We could alter this task by having the patient repeat the problem back to the tester as soon as it is read. This would more clearly isolate the attentional factor and give us direct information about how much the patient understood. Often such qualitative information is far more important to ultimate diagnostic inference than the score itself.

It is also common for patients who have tried unsuccessfully to solve a problem to claim that they could do better with pencil and paper. It is important to provide the pencil and paper and determine what effect this has on performance. Occasionally, a patient who is blocked by anxiety while attempting to solve the problem mentally will then solve the problem with great ease after writing only one number on the page. The reassurance of the paper and written number releases the attentional blocking and allows the patient to solve the problem easily. In this instance the extra observation derived from extending the limits of the test is crucial in understanding the initial failure.

Any test is a potential candidate for restructuring and testing its limits. One more example will suffice for our purposes. At the beginning of the testing session, I administered the Trail Making Test (Armitage, 1946) to a very depressed patient. He appeared to have no obvious difficulty on either part A or B, but his times were extremely slow. At the end of the 3-hour testing session, I repeated the Trail Making Test with the additional instruction to work as fast as possible without regard to error. There is, of course, the obvious practice effect in repeating this test so quickly, but this effect could not account for the enormous improvement the patient derived from more direct instruction and encouragement. His second performance was well within the normal range, indicating that is initial poor performance was the result of his depression.

## SUMMARY

If neuropsychologists are to use test results to make inferences abut the brain, they must be careful to exclude all nonbrain influences. In patients with both physical disabilities and secondary emotional problems, these other influences are many. In such situations the neuropsychologist cannot hide behind a fixed test battery and interpret scores in a rote, standardized way. Sensory and motor limitations will completely eliminate the possibility of administering many tests and compromise the usual interpretation of others. Attentional disturbances associated with anxiety or depression will often result in impaired test performances which are most impaired on the more sensitive neuropsychological tests. The challenge to neuropsychologists working with this population is to put aside cookbook interpretations based on fixed test batteries and become active, knowledgeable investigators.

Neuropsychologists must have a wide variety of tests and evaluation techniques to choose from when attempting to bypass sensory and motor limitations. They must constantly be forming hypotheses about the nature of underlying disabilities and testing these hypotheses with additional tests and techniques. They must be acutely aware of the often subtle effects of attentional disturbances on test results and differentiate these effects by making complex comparisons of one test to another and restructuring and extending tests whenever appropriate. Neuropsychologists need to develop a broad, conceptual understanding of what skills must be assessed in order to make accurate and meaningful inferences about brain function. Then they must set about the difficult task of assessing those skills despite interference from the limitations that the patient brings to the testing session.

## NOTE

The author wishes to thank Craig Van Dyke for his careful reading of this chapter and for his many helpful suggestions. The author also wishes to thank Rita Emelia for typing and secretarial support.

## REFERENCES

Anderson, R. & Sisco, R. (1977). *Standardization of the WISC-R Performance scale for deaf children.* Washington, D.C.: Gallaudet College Press.

Armitage, S. G. (1946). An analysis of certain psychological tests used for the evaluation of brain injury. *Psychological Monographs, 60,* 277.

Arthur, G. (1947). *A point scale of performance tests. Revised form II.* New York: Psychological Corp.

Bender, L. A. (1938). A visual motor Gestalt test and its clinical use. *American Orthopsychiatric Association, Research Monographs,* No. 3.

Bender, M. B. (1979). Defects in reversal of serial order of symbols. *Neuropsychologia, 17,* 125–138.

Benton, A. L. (1974). *The revised visual retention test* (4th ed.). New York: Psychological Corp.

Benton, A. L. & Hamsher, K. (1978). *Manual for the multilingual aphasia examination.* Iowa City: University of Iowa.

Berger, L., Bernstein, A., Klein, E., Cohen, J. & Lucas, G. (1964). Effects of aging and pathology on the factorial structure of intelligence. *Journal of Consulting and Clinical Psychology, 28,* 201–203.

Boller, F. & Vignolo, L. A. (1966).

Latent sensory aphasia in hemisphere-damaged patients: An experimental study with the Token Test. *Brain, 89,* 815–831.

Buschke, H. & Fuld, P. A. (1974). Evaluating storage, retention, and retrieval in disordered memory and learning. *Neurology, 11,* 1019–1025.

Davison, L. A. (1974). Current status of clinical neuropsychology. In R. M. Reitan & L. A. Davison (Eds.), *Introduction in clinical neuropsychology: Current status and applications.* Washington, D. C.: V. H. Winston & Sons.

De Renzi, E. & Vignolo, L. A. (1962). The Token Test: A sensitive test to detect disturbances in aphasics. *Brain, 85,* 665–678.

Diller, L., Ben-Yishay, Y., Gerstram, L. J., Goodkin, R., Gordon, W. & Weingerg, J. (1974). *Studies in cognition and rehabilitation in hemiplegia* (Rehabilitation Monograph, no. 50). New York: New York University Medical Center Institute of Rehabilitation Medicine.

Elliott, H., Glass, L. C. & Evans, J. W. (1987). *Mental health assessment of deaf clients.* Boston: Little, Brown/College Hill.

Golden, C. J., Hammeke, T. A. & Purisch, A. D. (1980). *Manual for the Luria-Nebraska Neuropsychological Battery.* Los Angeles: Western Psychological Services.

Gorham, D. R. (1956). A proverbs test for clinical and experimental use. *Psychological Reports, 1,* 1–12.

Gronwall, D.M.A. & Sampson, H. (1974). *The psychological effects of concussion.* Auckland: Auckland University Press/Oxford University Press.

Halstead, W. C. (1947). *Brain and intelligence.* Chicago: University of Chicago Press.

Heller, B. W. & Harris, R. I. (1987). Special considerations in the psychological assessment of hearing impaired persons. In B. W. Heller, L. M. Flohr & L. S. Zegans (Eds.), *Psychological interventions with sensorially disabled persons.* New York: Grune & Stratton.

Hirschenfang, S. (1960). A comparison of WAIS scores in hemiplegic patients with and without aphasia. *Journal of Clinical Psychology, 16,* 351.

James, W. (1958). *Talks to teachers.* New York: W. W. Norton.

Lashley, K. S. (1929). *Brain mechanisms and intelligence: A quantitative study of injuries to the brain.* Chicago: University of Chicago Press.

Leiter, R. G. (1969). *Examiner's manual for the Leiter International Performance Scale.* Chicago: Stoelting.

Lezak, M. (1983). *Neuropsychological assessment.* New York: Oxford University Press.

Luria, A. R. & Majovski, L. V. (1977). Basic approaches used in American and Soviet clinical neuropsychology. *American Psychologist, 32,* 959–968.

Porteus, S. D. (1959). *The Maze Test and clinical psychology.* Palo Alto, Calif.: Pacific Books.

Rapaport, D. (1945). *Diagnostic psychological testing* (Vol. 1). Chicago: Year Book Publishers.

Raven, J. C. (1960). *Guide to the standard progressive matrices.* London: H. K. Lewis.

Reitan, R. M. (1955a). An investigation of the validity of Halstead's measures of biological intelligence. *Archives of Neurology and Psychiatry, 73,* 28–35.

Reitan, R. M. (1955b). The relation

of the Trail Making Test to organic brain damage. *Journal of Consulting Psychology, 19,* 393–394.

Reitan, R. M. (1957). Differential patterns of results of lateralized and localized cerebral lesions. *Proceedings of the Fifteenth International Congress of Psychology,* 208–209.

Reitan, R. M. (1964). Psychological deficits resulting from cerebral lesions in man. In J. M. Warren & K. Akert (Eds.), *The frontal granular cortex and behavior.* New York: McGraw-Hill.

Rey, A. (1964). *L'examen clinique en psychologie.* Paris: Presses Universitaires de France.

Savage, R. D., Britton, P. G., Bolton, N. & Hall, E. H. (1973). *Intellectual functioning in the aged.* New York: Harper & Row.

Seashore, C. E., Lewis, D. & Saetveit, D.L.V. (1960). *Seashore measures of musical talents* (rev. ed.). New York: Psychological Corp.

Smith, A. (1968). The Symbolic Digit Modalities Test: A neuropsychologic test for economic screening of learning and other cerebral disorders. *Learning Disorders, 3,* 83–91.

Spiers, P. A. (1981). Have they come to praise Luria or to bury him? The Luria-Nebraska battery controversy. *Journal of Consulting and Clinical Psychology, 49*(3), 331–341.

Taylor, E. M. (1959). *The appraisal of children with cerebral deficits.* Cambridge: Harvard University Press.

Thurstone, L. L. (1938). Primary mental abilities. *Psychometric Monographs,* No. 1.

Thurstone, L. L. & Thurstone, T. G. (1962). *Primary mental abilities* (rev. ed.). Chicago: Science Research Associates.

*Webster's seventh new collegiate dictionary,* s.v. "attention."

Wechsler, D. (1945). A standardized memory scale for clinical use. *Journal of Psychology, 19,* 87–95.

Wechsler, D. (1955). *Wechsler Adult Intelligence Scale manual.* New York: Psychological Corp.

Wechsler, D. (1974). *WISC-R manual.* New York: Psychological Corp.

Wechsler, D. (1981). *WAIS-R manual.* New York: Psychological Corp.

Zieziula, F. R. (1982). *Assessment of hearing impaired people.* Washington, D.C.: Gallaudet College Press.

# PART III

## Treatment/Rehabilitation

After the assessment team has assembled and has discussed and analyzed all pertinent information about the patient, a treatment and rehabilitation plan is formulated. Since space does not permit a discussion of every disability, four chapters in this section focus on conditions which are paradigmatic of four major groups of disabling conditions. These include disabilities whose progressive or nonprogressive course begins at birth, in childhood, or in adulthood. In addition, a chapter is devoted to psychological aspects of facial disfigurement, an important but much-neglected area.

Nonprogressive disabilities with an onset at birth or in childhood include spina bifida and cerebral palsy. The prevalence of spina bifida varies between 1 to 2 per 1,000 births in the United States (Alter, 1962). Cerebral palsy is more common, with an incidence of between 4.3 to 4.6 per 1,000 (Nelson & Ellenberg, 1978; Orr-Kissner, 1978). As with other disabilities, psychosocial difficulties are often a significant impediment to successful treatment and rehabilitation. Employing a perspective informed by awareness of biological, developmental, psychological, and environmental factors, Gabriella Molnar describes how the early onset of a disabling condition profoundly affects the child's activities, sense of self, and interpersonal relationships, in addition to creating stress in the family. The potential for stress arises initially with the diagnosis and appears to traverse three stages: initial impact, immediate response, and gradual adaptation. Professionals can profoundly influence parental responses in these as well as later phases.

The professional must offer compassion as well as information and should neither communicate an overly pessimistic outlook nor raise false hopes. In addition, while professionals are essential, Molnar suggests that the leading role during treatment and rehabilitation must always remain with the family. Using clinical vignettes, Molnar describes common psychosocial responses of disabled children, including insecurity and timidity, low self-esteem, dependency, isolation, lack of self-control, and identity diffusion, among others. She finds severe depression to be rare. Paradoxically, children with mild disabilities often exhibit the most severe psychopathology, perhaps because they are expected to compete (and succeed) in an environment for which they are comparatively ill prepared. Preventive and remedial interventions should begin with the diagnosis and should continue with every professional contact. Molnar illustrates such interventions with clinical vignettes and ends with the case of a 3½-year-old girl with diplegic cerebral palsy, who successfully adapted to her disability.

Gloria Eng offers a brief history of Duchenne muscular dystrophy (DMD), a progressively deteriorative disability with onset in childhood, and then details its incidence (about 1 in 4,000 live male births [Dubowitz, 1978; D. S. Wood, pers. com., 1985]), prevalence (between 12,000 and 20,000 cases in the United States), genetics, clinical manifestations, and course. The story is not a happy one. DMD is often diagnosed at between 2 to 4 years, usually as a result of motoric abnormalities. There is progressive deterioration until a plateau occurs between the ages of 5 to 7. False hopes engendered during this period must be addressed, since they will inevitably be dashed. Most patients lose the capacity to walk between the ages of 8 to 12. Few live past their teens. The slow, relentless deterioration is especially difficult because of the age of the victim, management problems resulting from the increasingly pervasive character of the disability, the need to endure the daily routine of physical therapy, and the child's increasingly sharp awareness that there is something desperately wrong. The continual need to deal with the disability in terms of physical care and environmental adaptations also takes its toll on family members. Psychosocial interventions for patient and family are crucial. Group therapy is often useful. Eng suggests that treatment should emphasize the importance of functioning despite disability and that the professional's attitude should be positive, anticipatory, and supportive. After the age of 12, the life of the patient and family is continually punctuated by crises as the ravages of the disease become more prominent and life threatening. After the child's death, the professional must remain available to parents to assist in completing their grieving process.

Between 8,000 and 10,000 persons are diagnosed with multiple sclerosis (MS) each year, resulting in a prevalence of about 1 person per 1,000 at any given time in the United States (Baum & Rothschild, 1981; Poser, 1987). Sheldon Berrol describes the history, symptom manifestation, and characteristic emotional responses of the multiple sclerosis disease process. While certain emotional characteristics were reported by Charcot (1877) and others over 100 years ago, a lack of understanding of multiple sclerosis and its sequelae has obscured perceptions in the past. Berrol feels that only in the past few years has an informed understanding of these difficulties been attained. He forcefully demythologizes the "multiple sclerosis personality." There is none, he avers, except in the sense that any persons respond to a progressively disabling condition in common, specific ways based on their repertoire of coping skills, the quality of their interpersonal relationships, their ego strength, and other variables. This "syndrome" was often characterized as consisting of inappropriate euphoria, denial, and hysteria, though little research has demonstrated this to be the case.

Berrol suggests that, rather than being influenced by a single personality constellation, it is more likely that emotional reactivity is influenced by the uncertainty in diagnosis, the ebb and flow of the disease process, a lack of information about what to expect or when, and the increasing pressure to maintain one's current level of functioning despite decreased physical capabilities. Denial is often a very useful coping mechanism and should not be discounted (Lazarus and Folkman [1984] marshall empirical data to make a similar point). In addition, what is perceived as denial may often be the result of cognitive deficits, which often occur quite soon after the onset of the disease. Berrol suggests that much of the lore about emotional reactions in MS patients must be rethought, that patients' coping mechanisms should not be discounted—even when they are less effective than the professional might like. In addition, he stresses the importance of: information-sharing with patient and family; increased sensitivity on the part of the health care team to the sense of loss and powerlessness felt by the patient; integration of the patient into home and community as fully as possible; and the effectiveness of individual and group counseling in allowing patients to express anger, frustration, and depression. Increasing the functional capabilities of the individual through such means, he concludes, markedly reduces negative emotional reactions to the disease process.

Roberta Trieschmann describes the treatment and rehabilitation of individuals with spinal cord injury, the preeminent cause of nonprogressive adult-onset disability. About 10,000 persons a year suffer spinal cord injuries severe enough to render them paraplegic or quadriplegic. About 80% of the 200,000 spinal cord injury victims in the United States are young, active males, more than 60% of whom are between the ages of 15 and 29 (Trieschmann, 1988). The incidence of spinal cord injury is low, but its economic and emotional costs are very high. Trieschmann argues that there is no "spinal cord injury personality"—that there is no evidence, for example, of self-destructive tendencies among spinal cord injury victims. Instead, they are representative of a group in the population—adolescent and young adult males—which typically takes risks. Coping and adjustment mechanisms vary. They are specific to each individual and consistent with an individual's premorbid style. Often in the midst of planning a career, mapping out a life, these individuals must deal with the abrupt and profound life changes that their disabilities entail.

Trieschmann offers a social learning approach, as opposed to the medical model. Instead of assuming—as the medical model does—that organic treatments will promote adjustment unless hindered by personality problems, and therefore viewing the patient as a passive object to be "fixed," the social learning approach characterizes adjustment as a function of the interaction of multiple factors, including personal, organic, and environmental variables. While the medical model may be appropriate for the treatment phase, rehabilitation is a process in which the patient is expected to take an active part. Much work involves learning skills and strategies to readjust to new physical and environmental limitations. There is no evidence, Trieschmann states, for a stage theory of adjustment, nor does she see depression as a necessary precursor to adjustment. Moreover, she cautions, depression must be distinguished from ordinary unhappiness (a distinction well-known to Freud), just as denial—which can often be quite appropriate—must be distinguished from hope. Finally, treatment and rehabilitation should stress an individualized plan that enlists the participation of the patient to maximize function, independence, and community living.

The last chapter in this section discusses interventions for the psychosocial problems associated with facial disfigurement. This disability is usually not life threatening, though it often exterminates social life for the disabled victim. Viewing a disfigured face is particularly unpleasant for most of us, because it threatens, perhaps more than any other disability, our sense of bodily integrity. This is because the face is a particularly commanding and intimate source of interpersonal information (Ekman, 1982). It cannot be hidden, as most other disabled parts can; it is the most public part of the body. The face is particularly important in infancy and childhood. It is the vehicle for expression of emotion, mood, and other nonverbal as well as verbal communications and is the primary "organ" of interpersonal relations. In addition, it is the site of the sensory organs and the orifices for air, water, and food intake. Perhaps our innate aversion to disfigured faces, our lack of ability to empathize, explains not only why persons with disfigured faces are so often shunned but also why so little has been written about emotional reactions to this condition.

To begin his classic work on stigma and its management, Goffman (1963) cited a fictionalized but all-too-real letter from a facially disfigured adolescent to a columnist for the lovelorn. While the emotional reactions expressed may not be characteristic of persons with disabilities, they provide a moving illustration of the potential impact of severely disfiguring conditions.

Dear Miss Lonelyhearts—

I am sixteen years old now and I don't know what to do and would appreciate it if you could tell me what to do. When I was a little girl it was not so bad because I got used to the kids on the block makeing fun of me, but now I would like to have boyfriends like the other girls and go out on Saturday nites, but no boy will take me

# INTRODUCTION

because I was born without a nose—although I am a good dancer and have a nice shape and my father buys me good clothes.

I sit and look at myself all day and cry. I have a big hole in the middle of my face that scares people even myself so I cant blame the boys for not wanting to take me out. My mother loves me, but she crys terrible when she looks at me.

What did I do to deserve such a terrible bad fate? Even if I did do some bad things I didn't do any before I was a year old and I was born this way. I asked Papa and he says he doesnt know, but that maybe I did something in the other world before I was born or that maybe I was being punished for his sins. I dont believe that because he is a very nice man. Ought I commit suicide? (West, 1962, 14–15)

Norman Bernstein eloquently illustrates the proposition that the wounds occasioned by facial disfigurement are more than skin deep. Facial disfigurement does not hinder mobility, restrict sensation, or cause pain, yet it is a particularly difficult type of impairment, because it often results in great social isolation. The facially disfigured person may frighten other people, resulting in a "leper syndrome," low self-esteem, increased stress, and the like. The development of the person behind the facial mask is often markedly distorted if onset is at birth or in early childhood; later onset significantly changes previous psychological status and relationships. But for these individuals reactions vary markedly. For some persons, a relatively slight blemish may be extremely disturbing, while others may react less strongly to much more gross disfigurement. Psychological improvement may not correlate with patient satisfaction or dissatisfaction with results of medical treatment.

We often equate the face with the whole person. ("Let me show you my wife," a man may say, as he pulls out a small, flat, two-dimensional photograph of his wife's face.) Because of this and because the face is the site of social communication and expression of emotion, mood, and the like, an injury to the face may have even more impact than an impairment of another portion of the anatomy. Bernstein considers psychosocial interventions with facially disfigured persons fundamentally different from other forms of psychotherapy and rehabilitation. The battle is mainly with societal values and community reactions, with problems of symbolism and appearance, rather than with communication difficulties, sensory loss, pain, or lack of function or mobility. Chronic grief and the long-term nature of treatment and rehabilitation also differentiate interventions with these individuals from those with persons having many other impairments. Incidence figures for this condition—since it may arise from many different causes and is not tracked under a single category—are unavailable.

## REFERENCES

Alter, M. (1962). Anencephalus, hydrocephalus, and spina bifida. Epidemiology, with special reference to a survey in Charleston, S.C. *Archives of Neurology, 7,* 411–422.

Baum, H.M. & Rothschild, B.B. (1981). The incidence and prevalence of reported multiple sclerosis. *Annals of Neurology, 10,* 420–428.

Charcot, J.M. (1877). *Lectures on diseases of the nervous system.* London: New Sydenham Society.

Dubowitz, V. (1978). *Muscle disorders in childhood.* Philadelphia: W.B. Saunders.

Ekman, P. (Ed.). (1982). *Emotion in the human face* (rev. ed.). New York: Cambridge University Press.

Goffman, E. (1963). *Stigma: Notes on the management of spoiled identity.* Englewood Cliffs, N.J.: Prentice-Hall.

Lazarus, R.S. & Folkman, S. (1984). *Stress, appraisal, and coping.* New York: Springer.

Nelson, K.B. & Ellenberg, J.H. (1978). Epidemiology of cerebral palsy. In B. Schoenberg (Ed.), *Advances in neurology.* New York: Raven.

Orr-Kissner, C. (1978). *National significance project: Developmentally disabled individuals of normal intelligence.* United Cerebral Palsy Association of California.

Poser, C.M. (1987). Epidemiology and genetics of multiple sclerosis. In L.C. Scheinberg (Ed.), *Multiple sclerosis: A guide for patients and their families.* New York: Raven.

Trieschmann, R. (1988). *Spinal cord injuries: Psychological, social, and vocational rehabilitation.* New York: Demos Publications.

West, N. (1962). *Miss Lonelyhearts.* New York: New Directions.

# 5

# The Influence of Psychosocial Factors on Personality Development and Emotional Health in Children with Cerebral Palsy and Spina Bifida

*Gabriella E. Molnar*

Physical capabilities, cognitive function, and emotional health are the three principal factors that influence the prognosis and ultimate accomplishments of children with cerebral palsy or spina bifida. Long-term studies indicate that psychosocial disability is a significant source of the failure to attain physical independence and career expectations when these children reach adulthood (Cruickshank, 1978; Easton, 1985). These observations demonstrate the importance of emotional and personality development for long-range life achievements and functioning in society.

The biologic, psychologic, and social forces that shape the personality and behavior of children with cerebral palsy or spina bifida during the formative years tend to be different from those that operate in the lives of their able-bodied peers (Klapper & Birch, 1966; Parker, 1971; Wortis & Cooper, 1957). Despite these differences many children emerge as mature, self-sufficient adults. Yet, the variance in life experience creates a risk for psychosocial dysfunction, especially with physical disabilities originating at birth or early childhood (Minde et al., 1972).

This discussion will examine the interactive cycle of altered biological circumstances and psychosocial environment and will highlight preventive and remedial strategies for alleviating the risk of maladjustment.

## INTERACTION OF BIOLOGIC FACTORS AND PERSONALITY DEVELOPMENT

The study of personality development has generated a number of conceptual schemata in the behavioral sciences, including psychoanalytic theory (Freud, 1965), behaviorism (Skinner, 1974), social learning theory (Sussman, 1965), and temperament theory (Buss & Plomin, 1975). Advances in the neurosciences stimulated the emergence of a psychobiological viewpoint of behavior and emotions (Gianotti, 1979; Mueller, 1983). Continuing research should elucidate many additional, currently unknown aspects of the neurobiology of affect, temperament, and personality. These considerations provide a comparative framework for exploring the difficulties handicapped infants and children may encounter and for understanding the origins of potential deviance.

### Physical and Associated Disabilities

One view of early personality development, Mahler's (1963) concept of individuation and separation, has particular relevance for children with congenital or perinatally acquired disability. The prenatal biologic symbiosis is replaced at birth by a psychosocial symbiosis. As the infant's perceptual, motor, and cognitive abilities develop, the concept of separateness emerges, and the development of creeping and walking enables the infant to take active control of separation. The parent-infant unit becomes increasingly individualized as the acquisition of language provides an additional tool of control. This process, which ordinarily occurs from 6 to 36 months, is delayed or curtailed in children with cerebral palsy or spina bifida. When the stage of active physical separation is not achieved, prolonged dependence and the inability to take control can hinder the normal developmental thrust toward individuation.

Erikson (1963) views the process of personality development in terms of crises that are characteristic of each age; the resolution of these crises is the task of each developmental stage. Autonomy versus shame and doubt are issues in the second year of life that are resolved with the achievement of locomotion and sphincter control. How does a child with spina bifida resolve these developmental issues in the face of neurogenic sphincter dysfunction and inability to walk? The critical issue in later childhood is industry versus inferiority, which is also difficult to resolve for a child who is dependent in many aspects of daily life. Identity versus identity diffusion, concern with how one appears in the eyes of others, and increasing interest in the opposite sex are the critical issues of adolescence, often a time of emotional turmoil whether the young person is handicapped or not (Abramson et al., 1979; Dorner, 1976). Physical disability creates a visible stigma, which Goffman (1963) called "spoiled identity." Physical appearance and the establishment of sexual identity are serious concerns for disabled adolescents, particularly those with spina bifida, who have neurogenic sexual dysfunction (Anderson, 1979). Failure to resolve these critical aspects of identity and self-image may precipitate emotional disturbances in disabled adolescents (Magill & Hurlbut, 1986; Minde, 1978). The question one must raise is not why do some children with cerebral palsy or spina bifida develop emotional difficulties but why is it that so many of them do not. We must give credit to the adaptability of human nature and, especially, to the children and their families who find the inner strength to cope physically and emotionally.

Sensory and perceptual deficits, as well as body image concerns, can further complicate the development of identity, since, for a youngster with spina bifida who has a thoracic neurosegmental lesion, complete paralysis, and no sensation in the legs, these issues differ considerably from those of other children. A 5-year-old, bright, articulate child with low thoracic lesion for whom standing and ambulation with crutches and orthosis was planned, did not want to comply with this program. His lack of cooperation was attributed to an emotional problem. Discussing the reasons for his behavior, he explained that he was afraid because he felt like he was floating in midair with no stable ground. Impaired vision or hearing and visual-motor or spatial perceptual deficits associated with cerebral palsy can make interpretation of the surrounding world difficult (Birch, 1964). A young child with such deficits may in new situations show excessively frightened behavior, which can be misconstrued as temper tantrums.

Personality, behavior, and affective disorders may be the manifestations of an organic brain dysfunction (Heilman & Valenstein, 1984). A well-known example is the hyperactive, disinhibited behavior, short attention span, and distractibility seen in some children with cerebral palsy (Denhoff & Robinault, 1960). A hyperverbal behavior usually referred to as "cocktail party personality" was described in hydrocephalic children with spina bifida. Excessive chatter, often with irrelevant contents, and a discrepancy between deceivingly good verbal ability and poor language function are characteristic of this syndrome (Swisher & Pinscher, 1981).

Although the roles of the limbic lobe and reticular activating system have been known for some time, investigations in the neurosciences continue to elucidate other neuroanatomic and physiologic correlates of emotion and behavior (Mueller, 1983). Application of these data to children with central nervous system deficits is inferential since there are no direct studies available on this population. Nevertheless, it is intriguing to speculate that deviant behavior, for example, the passive dependency, depression, and social withdrawal observed among some children and adults with cerebral palsy, may have dual roots of environmental and biologic origin.

Quite recently recognized is the association of personality changes, hypomanic-depressive states, and aprosody with right hemispheric lesions (Davidson, 1983; Galaburda et al., 1978; Springer & Deutsch, 1981). Originally described in adults after cerebrovascular accidents, these symptoms were also observed in children with acquired left hemiparesis.

The history of a boy with left hemiparetic cerebral palsy suggests the possibility of similar etiology for severe maladjustment. He was born at term after prolonged labor but had no neonatal complications. Right hand preference was recognized around 6 months of age, and the diagnosis of left hemiparesis was made. He had slight delay of gross motor milestones but walked around 18 months of age; he had a mildly abnormal gait and used his left hand quite well for assistive function; he had no cortical sensory deficit, perceptual dysfunction, hemianopsia, or clinical or EEG evidence of seizure disorder. His IQ was 104. He was the thirdborn of four children, with one older brother and an older and younger sister; his father was a construction worker, his mother a homemaker. The parents went through a slightly prolonged mourning process upon the diagnosis of cerebral palsy but responded well to supportive counseling.

This was a warm, close-knit family; the boy's mild disability made the family ties even stronger. Both parents were hard working and affectionate people. All children were expected to perform appropriate chores around the home, and this expectation was enforced in a matter-of-fact style without pressure. The boy was exuberant; he was a tease and a clown, perhaps too much so. He had no behavior problems and was placed in a regular kindergarten and primary school. There were no apparent social adjustment problems, and he had many friends at school and in the neighborhood. Scholastically, however, he was an underachiever. His parents and teachers said he did not seem to care. There were many conferences with the family, school, and the rehabilitation team. The boy and his parents had regular counseling, and his school performance improved.

Nevertheless, over the years there was a gradual change in his personality and behavior. Initially jocular and outgoing, by the age of 12 he had become quite subdued, withdrawn, and depressed. His face reflected very little affect, and his speech had an aprosodic quality. With psychiatric treatment there was a temporary resolution of school problems and depression. By age 14, however, his difficulties once again reached a crisis stage, but he refused all medical visits and treatment. At age 16 he dropped out of school and rejected all attempts of help—which he considered an interference with his life. Through his mother, who kept in touch with us, we learned that he eventually found work with the sanitation department, lived at home, and remained withdrawn and depressed.

This boy had many assets and should have succeeded in life. He had a cosmetic rather than functional handicap, and he possessed normal intelligence. The family constellation was favorable and supportive. No early signs augured failure, though, perhaps in retrospect, his excessive jocularity should have been considered a warning sign. The clinical findings and course of this case suggest the possibility of an affective disorder associated with damage to the right hemisphere. The hypomanic behavior and depression, the indifference to school failure, and the striking change and loss of affect in facial expression and speech resemble the combination of signs attributed to nondominant hemispheric lesion.

## INFLUENCE OF PSYCHOSOCIAL FACTORS

Emotional and personality development is influenced by the child's interactive experiences with family, peers, professionals, and society at large (Bergsma & Pulver, 1976).

### Family

The birth of a handicapped child is a time when parents are most vulnerable and in need of intensive support to adjust (Green & Solnit, 1964). Complications of the birth process and neonatal period are frequent among infants with cerebral palsy or spina bifida. Newborns with hypoxic encephalopathy are often too sick to be cuddled and fed. Infants born with spina bifida have surgery within the first few days of life to close the cystic myelomeningocele sac, and the majority need a shunt insertion for hydrocephalus (Badell-Ribera, 1985). Neonatal complications may delay or distort the usual early parent-infant attachment and bonding—a concept borrowed from ethology (Klaus & Kennel, 1976). In some cases, these events may lead to disorders of parenting.

The dynamics of adjustment to parenting a handicapped child and the long, often painful road that leads to effective coping are complex (Freeston, 1971). "Acceptance" would be an erroneous choice of word in this context. One mother, an exceptionally warm person who was instinctively creative in nurturing her child's development, expressed her feelings by saying: "When your child is handicapped, it is like living under a cloudy sky. The sun never shines." Grieving is a universal response in the process of adjustment (Solnit & Stark, 1961), but each family has its own coping style related to the parents' personalities and backgrounds. Parents with good ego strength are able to cope better than those with weak or neurotic personality traits. Their values and aspirations will also influence the process of adjustment (Kessler, 1977; Pearlman & Scott, 1981).

Having a handicapped child creates stress in the family unit (Chess et al., 1980; Dorner, 1973, 1975). The triadic interrelationships of mother, father, and child may be skewed. By necessity or because of her own emotional makeup, the mother may devote most of her attention to the handicapped child. An emotional alienation between the spouses may ensue. It has been suggested that there is a higher rate of separation and divorce in families of handicapped children; however, dissolution of the family usually occurs in marriages already stressed before the birth of the disabled child (Tew et al., 1974). In addition social isolation of the whole family has been described.

The family's child-rearing style has a profound effect on the development of the child's personality (Friedman & Friedman, 1977). Bringing up a handicapped child is the greatest challenge any parent can face (Wasserman et al., 1985). There are norms and expectations to guide the parents of nonhandicapped children, but what standards apply to a disabled child? (Shere & Kastenbaum, 1966; Shere, 1971). To their credit, most families have a natural understanding of their child's difficulties and instinctively adapt usual child-rearing practices (Katz, 1961; Kogan et al., 1974; Sears et al., 1957). Some parents may have difficulties in finding a healthy balance between demanding too much or too little and need intensive guidance and emotional support.

Denial, excessive anxiety, guilt, overprotection, and rejecting attitudes are commonly described pathologic reactions. However, both clinical experience and research data show that most parents make appropriate adjustments to meet the practical needs of their child (Tew & Laurence, 1973). Studies of handicapped children and their families by Hewett (1970), Barsch (1986), and McMichael (1971) showed that although pathologic coping and parenting do occur, they are less frequent than generally assumed. The investigators concluded that the stereotyped image of parents of handicapped children as guilt ridden, overanxious, overprotective, or rejecting is not justified and that each family must be judged individually and without preconceived notions.

**Professionals**

Medical and allied health professionals play only a peripheral role in the lives of most families with able-bodied children. This is not so when a child is handicapped: there will be many visits and meetings with different professionals while the child is growing up. The first encounter is at the time of the child's birth or when a disability is recognized (D'Arcy, 1968). Informing parents about the diagnosis of disability and its complications is similar to crisis intervention (Milunski, 1981). For parents, the reactive process set into motion by this information has three stages: initial impact, immediate response, and gradual adaptation. The physician's communication skills play a significant role in all three phases (Korsch & Negrete, 1972; Wolraich, 1982). Unfortunately, classes in communication skills are not included in the traditional medical educational curricula despite the fact that conveying the diagnosis of a disability or grave illness is one of the more difficult tasks physicians must perform. For example, there are different ways one can provide the same factual information to parents (Taft et al., 1983). A detached discussion that includes only the negative consequences of the disability and the limitations imposed by it can have a devastating impact. It can have a lasting influence on parents' attitudes and behavior toward their child. Undoing the effect of such an interview may require considerable time and effort.

While it is necessary to give the family an honest account of the anticipated functional limitations, it is also essential to balance this information by emphasizing the functional potentials and abilities that are preserved. The message that all is not hopeless must be conveyed in a compassionate, caring manner and supported by information about the rehabilitation process that lies ahead. In cerebral palsy and spina bifida the multifactorial circumstances lead to many uncertainties of prognostication (Badell-Ribera, 1985; Molnar, 1979). Therefore, one must be careful to create neither false hopes nor an overly pessimistic outlook at the initial informing interview. There, professionals should share considerations with the parents, acknowledge the hardship of waiting for more definite predictions, and reassure them that all expectations and concerns will be candidly discussed as they become appropriate.

The rehabilitation process is a joint undertaking by the family and the professional team, but the leading role in this partnership must remain with the family (McKeith, 1976; Meadow & Meadow, 1971). The team must make it clear that it is there to advise and help but not to take over the upbringing of the child (Beckwith, 1976). There are a number of pitfalls that should be avoided in the relationship between families and the professional team. Conflicts may arise about realistic expectations and functional goals. Some parents place an excessive emphasis on treating the physical dysfunction and other deficits associated with the disability at the expense of the child's developmental, educational, and social needs. Sometimes this reflects the attitude of professionals working with the child, and other times, the family's interpretation of that attitude. These issues must be resolved by repeatedly counseling the parents about priorities for the child (Kolin et al., 1971). The team must not shun its responsibilities for advocating the long-range interest and emotional welfare of the child, a task that is often more demanding than agreeing with parents' requests for prescribing more physical therapy or humoring their unrealistic expectations. Excessive dependence by parents on the rehabilitation team may originate from the parents' feeling of being overwhelmed by the demands of bringing up a handicapped child or from their lack of self-confidence to fulfill this task. The team must be sensitive to these feelings and avoid an overbearing, authoritarian approach that can undermine their assurance of themselves as competent parents (McKeith, 1973). Erosion of the parental role is a paradoxical outcome of the rehabilitation process, for a successful intervention should lead to strong self-reliance in the child and family.

**Peers and Society**

Social interaction with disabled persons is complicated by uncertainties of able-bodied people about the applicability of rules of behavior (Sussman, 1965). There are no standards for interacting with a disabled person that able-bodied people could model as they learn the conventions of social behavior (McDaniel, 1976; Perlman & Routh, 1980). A sociologic concept proposed by Richardson (1970, 1971; Richardson et al., 1961, 1974) suggests that handicap is interpreted as a deviation from social standard, and this in turn creates a "violation of expectations." Failing to conform to the norms of their peer group and society at large, handicapped children may face difficulties gaining social acceptance (Osman & Blinder, 1982; Richardson, 1969).

A lovely and determined young lady whom I had the privilege to follow and watch as she grew up wrote: Emotionally, the cerebral palsied person encounters many difficulties. He has to accept the fact that people see him as different and thus he is also treated differently. The inner struggles that he has are harder to tackle since they deal with the feelings he has toward his situation as well as his need to accept the reactions of people around him (Ortigas, 1983, 327).

Physical access has been a major problem for disabled people in contemporary society. Although there are continuing attempts to eliminate architectural and other environmental barriers, geographic integration does not necessarily ensure social integration. Despite Public Law 94-142, which mandates a least-restrictive environment for the education of the handicapped, and despite other federal and state laws to promote equal opportunities for the disabled, these promises remain unmet.

## PSYCHOSOCIAL FUNCTION AND DYSFUNCTION

"One of the hardest problems a person with cerebral palsy has to cope with is low self-esteem." He may find it difficult to accept "that, despite his handicap, he can stand on his own and become a useful member of society. This is a very hard thing to do and requires a lot of courage and willingness to go through all the hardships to prove that he is not given any sort of special treatment just because he has a handicap" (Ortigas, 1983, 327). These pensive words summarize the personal experience of a college student and the doubts and obstacles that a disabled child must overcome (Reynell, 1973; Rubinstein, 1982).

Difficulties with interpersonal relationships often arise from limited opportunities to learn and practice social skills (Strain & Kerr, 1979). Timid, insecure behavior is one manifestation of poor self-concept (Teplin et al., 1981; Van Putte, 1979), while compensatory manners of behavior such as clowning or verbal aggressiveness represent the other end of the scale. Lack of self-confidence can interfere with developing a trust in others and with the establishment of lasting friendships and attachments.

Emotional and personality problems extend over a spectrum from minor deviations to severe pathology (Freeman, 1967; Friedman, 1981; Fulthorpe, 1974; Oswin, 1967; Seidel et al., 1979). Ironically, children with mild disabilities seem to be more prone to maladjustment, perhaps because they are expected to compete in a society for which they are ill-prepared from their childhood experiences.

Confusion about the child's capabilities or low expectations set by the family, teachers, and other significant persons undoubtedly contribute to the immature personality, unnecessarily dependent attitude, and passive behavior observed in some handicapped children (Shakespeare, 1975; Wasserman, 1986). However, placing the responsibility on conditioned interaction would be an oversimplification. The psychodynamics of personality development are also influenced by the innate emotional and temperamental characteristics of a child and possibly by the nature of the damage to the central nervous system. Moreover, it is not inconceivable that an anoxic insult that affected the motor structures may also cause damage to the reticular activating system and play a role in the passivity, lack of drive, and poor motivation displayed by some children with cerebral palsy.

By necessity or by misguided good intentions, disabled children receive a greater share of attention from adults, including family, friends, teachers, and professional persons, than their able-bodied peers (Easton, 1985; Furgang & Yerxa, 1979; Ispa, 1981). In the course of rehabilitation and other modes of remedial intervention, the focus of attention revolves around the disabled child. An undesirable by-product is that the very same process that is designed to promote independence may foster an egocentric personality and manipulative, attention-getting behavior. The child who was taught and is able to perform certain self-care skills in therapy but does not perform them at home or in school, is well known to all who are involved in the rehabilitation of disabled children.

Socially unacceptable, uncontrolled behavior may be an expression of frustration. Inability to move around, to explore, and to take active control because of physical limitations may be a precipitating factor of temper tantrums in young, disabled children during the motorically driven stage of development. Enforced immobilization imposed after orthopedic surgical procedures and during hospitalization is an emotionally trying time for the preschool child (Tew & Laurence, 1976). Inability to communicate is another source of frustration and may be a cause of "acting out" behavior in children with athetoid or spastic quadriparetic cerebral palsy, good language function, yet impaired control of speech production. Such behavior deviations often yield to practical means of remediation such as the use of nonverbal communication devices in case of a severe speech impediment (Vanderheiden & Grilley, 1983).

The adolescent quest for identity is complicated by anxiety over appearance and acceptance by peers and by unresolved issues of sexuality (Graham, 1985; Jessell & Veltri, 1986). Rebellion against authority can take many forms, including noncompliance with medical and physical maintenance regimens. At this stage, when rapid growth threatens deterioration of musculoskeletal status, such regimens may become increasingly necessary. When independent social life becomes the accepted standard, the disabled adolescent's need to rely on others can magnify already low self-esteem and lead to self-imposed social withdrawal and isolation (Freeman, 1970).

Severe depression, reactive or perhaps organic, as in the earlier example, does not occur often. Generally it represents a culmination of psychoemotional pathology that preceded adolescence (Milavic, 1985). "Catastrophic reactions" are similarly rare. This term, used by Crothers and Paine (1961) in citing Goldstein, describes dramatic cases of flight from intractable reality, when individuals have to face situations to which they are not equal. The example of a 12-year-old girl with spastic diplegic cerebral palsy may illustrate such a case, this with an unusual outcome. She was born prematurely, the second child of a single career woman of good financial means. The girl's older brother was a varsity athlete and excelled scholastically. On the first visit, at age 5, the mother came carrying the child, who was capable of walking independently. She explained that this way people did not notice, stare, or ask embarrassing questions about the child's peculiar gait. After much counseling, the mother seemed to come to better terms with her daughter's disability. The child underwent surgery after which her walk improved. Her IQ was 120 and she entered regular school. Both mother and child seemed to do well emotionally, and there were no school problems. From 8 to 10 years, contacts were sporadic because of the mother's career obligations. At 12 years, returning for a visit after a 2-year absence, she was a good-looking preadolescent. Her school report indicated average achievements. However, she seemed insecure in social contact and was noncommunicative in the presence of her mother. Asking how she was doing opened a floodgate of complaints from her mother about the girl's horrible looks, daydreaming, laziness, lack of cooperation, poor school performance, and inferior social graces. For the entire time, the girl sat with an emotionless face, ignoring what was said and rocking herself while softly humming a tune.

After I requested the mother to leave the room so that her daughter could be interviewed alone, the girl's attitude changed markedly. She became animated and related stories about her successes in dancing, sports, social life with friends, and school work. However, from discussion with her grandparents, these achievements proved to be imaginary. The revealed history was that over the years the girl began to tell her tales with increasing frequency at home, at school, and to friends and strangers. When confronted with the truth, she did not deny she had made up her stories. She became known as a pathologic liar, to the frustration of her family and friends. Reality testing on psychiatric evaluations showed no delusional ideation. The opinion was that her pathologic escape from reality into a make-believe world required psychiatric treatment. After a brief course of therapy, she refused further treatment. As her storytelling continued, she became a social embarrassment. She was withdrawn from school and obtained her high school diploma by having a home tutor. She did not enter college and at the age of 19 was a social recluse, except for a companion retained by her well-to-do family. It would be perilous to speculate what initiated and perpetuated this girl's intense need to exchange real life for an imaginary world. Perhaps continued, intensive psychotherapy would have been able to clarify the contribution of her mother's barely disguised intolerance for imperfection, or assess the role of the girl's own personality in the development of this unusual behavior.

## PREVENTIVE AND REMEDIAL STRATEGIES

Anticipatory psychosocial intervention begins through the parents at the first informing interview and should continue at every subsequent encounter with the physician. The social worker, psychologist, nurse, and other members of the rehabilitation team play an essential role in this endeavor (Buscaglia, 1975; McDonald, 1962). Relating factual information and supportive counseling are two inseparable phases of interaction between parents and the physician. Learning about the negative aspects of disability is a painful experience that should be tempered by discussing the child's assets. With the parents' agreement grandparents, siblings, or other relatives should be involved in informational and supportive counseling (Crocker, 1981).

Because of their paramount role in nurturing their child's affective and personality growth, the parents must be made aware early that a psychosocial dysfunction may hamper optimal development and that they are influential in preventing it (Birenbaum, 1973). Being consistent with expectations, setting limits, allowing choices, and facing the consequences of one's actions are basic principles of child rearing that are equally applicable to disabled children. Success in achieving new skills, a goal shared by the parents and the rehabilitation team, serves to maintain the child's motivation and to establish the feeling of identity. However, short-term goals for attaining a specific motor skill should not conflict with the long-term aim of social and emotional maturity (Perrin et al., 1972). An undue emphasis on the treatment of physical disability that deprives the child of play, socialization, and other experiences of learning and pleasure is a misdirected priority that both parents and professionals need to realize and avoid.

Participation in a family support group is a valuable resource for parents (Birenbaum, 1973). Sharing common experiences is often a more effective and powerful influence than any professional advice or intensive counseling. Parents at different stages of the adjustment process usually derive considerable support from meeting other parents and benefit from the model of those who achieved successful parenting.

Each stage of development creates new interests and demands that a well-conceived rehabilitation plan should fulfill (Lemkau, 1961). The desire for mobility in early childhood is replaced by intellectual and social pursuits in late school age and adolescence. Parents and teachers often report a spurt of learning in young children with severe physical disabilities, for whom a wheelchair provides newly discovered independence and the means to take control (Breed & Ibler, 1982; Butler, 1986). An extreme example was a five-year-old child with spina bifida, congenital heart disease, and mental retardation. He had complete paralysis of the legs and good upper-extremity strength and function. When first brought to our office, he was maintained in a travel chair, a modified wheelchair device that is not self-propelant. It became immediately evident that his most distressing pathology was a severe behavior problem. The child screamed almost incessantly; he slammed his head against the high back of his chair. His mother stated that this was his usual behavior and that he had fallen out of his chair several times when it toppled over from his violent head banging. We decided to try seating him in a conventional wheelchair, which is more stable and provides self-controlled mobility, and to eliminate the high chair back, which only reinforced his self-stimulatory and self-abusive behavior. To our surprise, when he was placed in the new wheelchair, he briefly explored it, and on discovering how it worked he began to move around the room and down the hallway. Screaming and head banging ceased. Subsequently, a structured behavior modification program was introduced in which his wheelchair was one of the reinforcing stimuli. Although the child remained severely handicapped, his behavior was brought under satisfactory control, and he learned some basic adaptive skills commensurate with his mental abilities. On the other hand, the need to consider changing priorities during adolescence, for example, is well illustrated by a panel discussion of young disabled adults who expressed their resentment about missing classes before an important test in high school, because a session to practice walking was scheduled at the same time (Richardson, 1972).

The rehabilitation process should proceed on two simultaneous levels (Goldenson, 1978). Intertwined with considerations about promoting maximal motor function, integrity of the musculoskeletal system, and physical achievements must be the monitoring of cognitive and emotional development and other aspects of mental and social functioning (Molnar, 1981, 1985b). A useful method for exploring the home and social milieu is to review the child's schedule and activities on a typical weekday and weekend. Information gathered from these discussions may give many clues about family interaction, life-style, and values and about the child's physical and psychosocial adjustment, experiences, behavior, and interests that the traditional physical examination and the artificial environment of an office or clinic often fail to reveal. A perceptive examiner will be alerted to potential or evident areas of difficulty and will initiate anticipatory or remedial intervention.

An important phase of guidance is to encourage social encounters outside the home and to assist the family in seeking such opportunities. Recreation groups, camping, and adaptive sport events were found to promote both physical independence and social maturity of disabled children and adolescents (Bodzioch et al., 1986; Guttman, 1976; Molnar, 1985a). Direct teaching of social and "survival" skills and practice in simulated situations may be needed (Easton, 1985).

When counseling children and adolescents about sexual issues, parents should be consulted and their attitudes respected. Sexual counseling should be medically accurate and present all options realistically. The implications of sexual dysfunction associated with spina bifida must be explained simply, without medical terminology. The distinction between sexuality versus attachment and intimacy should be made clear (Diamond, 1974; Robinault, 1978).

Significant or severe emotional disturbance or psychosocial pathology requires formal psychiatric intervention (Freeman, 1970). In children, the need for psychotropic drug treatment arises most often for the control of organic hyperkinetic behavior disorder.

## EPILOGUE

Health professionals tend to be preoccupied with pathology. At this juncture, it behooves us to direct our attention to the countless children who with the help of their nurturing families have successfully met the challenge of disability (Cohen & Kohn, 1979). It is with this in mind that the following case history of a girl with diplegic cerebral palsy is related.

She was late in all motor milestones and achieved walking at 3½ years of age. She was an independent community ambulator despite an obviously abnormal gait. She was a good-looking, bright girl. Her father owned a retail shop and her mother was a homemaker; she had a younger brother. The family had no relatives in the area to provide support. The girl usually came to my office with both parents, who were warm, perceptive persons. They realistically acknowledged their daughter's disability but remained appropriately optimistic about her future. The girl was a well-motivated, determined person from early age on. She attended an integrated kindergarten and did very well. As usual, there was some anxiety when she started first grade, which was in another school, where she was the only disabled child in her class. She did well scholastically, but in the beginning she had to fend for herself to be accepted. There was one classmate who, in particular, teased her and on one occasion remarked that "You can't even go up the stairs like other kids." To this she replied, "But I can sing better than you, and I am better in math and reading than you are." As it turned out, this was a winning stroke; the girl never teased her again. (In fact they became "best friends", and their friendship continued into adulthood.) This 6-year-old girl was prepared to assert herself as equal to anyone. Her parents were able to instill in her a strong feeling of identity and positive self-image. She remained a well-adjusted child through the years of growing up and eventually achieved professional success and a happy marriage.

Two points merit comment. First, the foundations of emotional health are usually laid down in the early years. Second, while there is much research on pathologic adjustment, on what went wrong and why, few studies have explored the circumstances and events that make for success. It seems appropriate to close this discussion with advice from another successful young lady (Ortigas, 1983, 328). "One of the most valuable considerations we can give a person with cerebral palsy, or any handicap for that matter, is not to see him as a "special person," but as a person with special needs and help him meet these needs. Once in a while, a handicapped person should also be given the satisfaction of being useful and needed, of being the giver and not the receiver."

## REFERENCES

Abramson, M., Ash, M. J. & Nash, W. R. (1979). Handicapped adolescents—A time for reflection. *Adolescence, 14,* 557–565.

Anderson, E. (1979). The psychological and social adjustment of adolescents with cerebral palsy or spina bifida and hydrocephalus. *International Journal of Rehabilitation Research, 2,* 245–247.

Badell-Ribera, A. (1985). Myelodysplasia. In G.E. Molnar (Ed.), *Pediatric Rehabilitation.* Baltimore: Williams & Wilkins.

Barsch, R. H. (1986). *The parent of the handicapped child. The study of child-rearing practices.* Springfield, Ill.: Charles C. Thomas.

Beckwith, L. (1976). Caregiver-infant interaction as a focus for therapeutic intervention with human infants. In R.N. Walsh & W.T. Greenough (Eds.), *Environments as therapy for brain dysfunction. Advances in behavioral biology* (Vol. 17). New York: Plenum.

Bergsma, D. & Pulver, A. (1976). *Developmental disabilities: Psychological and social implications.* New York: Alan R. Liss.

Birch, H. G. (Ed.). (1964). *Brain damage in children: Biologic and social aspects.* Baltimore: Williams & Wilkins.

Birenbaum, A. (1973). Family management of cerebral palsy. *Pediatric Annals, 2*(12), 57–67.

Bodzioch, J., Roach, J. W. & Schkade, J. (1986). Promoting independence in adolescent paraplegics: A 2-week "camping" experience. *Journal of Pediatric Orthopedics, 6,* 198–201.

Breed, A. L. & Ibler, T. (1982). The motorized wheelchair: new freedom, new responsibility and new problems. *Developmental Medicine and Child Neurology, 24,* 366–371.

Buscaglia, L. F. (1975). *The disabled and their parents: A counseling challenge.* Thorofare, N.J.: C.B. Slack.

Buss, A. M. & Plomin, R. (1975). *A temperament theory of personality.* New York: John Wiley & Sons.

Butler, C. (1986). Effects of powered mobility on self-initiated behaviors of very young children with locomotor disability. *Developmental Medicine and Child Neurology, 28*(3), 325–331.

Chess, S., Fernandez, P. & Korn, S. (1980). The handicapped child and his family: Consonance and dissonance. *Journal of the American Academy of Child Psychiatry, 19,* 56–67.

Cohen, P. & Kohn, J. (1979). Follow-up study of patients with cerebral palsy. *Western Journal of Medicine, 130,* 6–11.

Crocker, A. C. (1981). Involvement of siblings of children with handicaps. In A. Milunski (Ed.), *Coping with crisis and handicap.* New York: Plenum.

Crothers, B. & Paine, R. S. (1961). *The natural history of cerebral palsy.* Cambridge: Harvard University Press.

Cruickshank, W. M. (1978). *Cerebral palsy, a developmental disability.* Syracuse, N.Y.: Syracuse University Press.

D'Arcy, E. (1968). Congenital defects: Mother's reaction to first information. *British Medical Journal, 3,* 796.

Davidson, R. J. (1983). Affect, repression and cerebral asymmetry. In L. Temoshok, C. Van Dyke, L. S. Zegans (Eds.), *Emotions in health and illness.* New York: Grune & Stratton.

Denhoff, E. & Robinault, I. P. (1960). *Cerebral palsy and related disorders.* New York: McGraw-Hill.

Diamond, D. (1974). Sexuality and the handicapped. *Rehabilitation Literature, 35,* 34.

Dorner, S. (1973). Psychological and social problems of families of adolescent spina bifida patients: A preliminary report. *Developmental Medicine and Child Neurology (Supplement), 29,* 24.

Dorner, S. (1975). The relationship of physical handicap to stress in families with an adolescent with spina bifida. *Developmental Medicine and Child Neurology, 17,* 755–775.

Dorner, S. (1976). Adolescents with spina bifida: How they see their situation. *Archives of Disease in Childhood, 51,* 439.

Easton, J.K.M. (1985). Psychosocial issues. In G. E. Molnar (Ed.), *Pediatric rehabilitation.* Baltimore: Williams & Wilkins, 99–120.

Erikson, E. H. (1963). *Childhood and society.* New York: W. W. Norton.

Freeman, R. D. (1967). Emotional reactions of handicapped children. *Rehabilitation Literature, 28,* 274.

Freeman, R. D. (1970). Psychiatric problems in adolescents with cerebral palsy. *Developmental Medicine and Child Neurology, 12*(1), 64–70.

Freeston, B. M. (1971). An inquiry into the effect of a spina bifida child upon family life. *Developmental Medicine and Child Neurology, 13,* 456.

Friedman, A. & Friedman, D. (1977). Parenting: A developmental process. *Pediatric Annals, 6,* 9.

Friedman, D. B. (1981). Developmental disabilities: Intervention strategies in the affective domain. In M. Lewis & L.T. Taft (Eds.), *Developmental disabilities: Theory assessment and intervention.* New York: Spectrum Publications.

Freud, A. (1965). *Normality and pathology in childhood: assessment of development.* New York: International Universities Press.

Fulthorpe, D. (1974). Spina bifida: Some psychologic aspects. *Special Education, 1,* 17.

Furgang, N. T. & Yerxa, E. J. (1979). Expectations of teachers for handicapped and normal first grade students. *American Journal of Occupational Therapy, 33,* 697.

Galaburda, A., Kemper, T., Le May, M. & Geschwind, N. (1978). Right-left asymmetries in the brain. *Science, 199,* 852–856.

Gianotti, G. (1979). The relationship between emotions and cerebral dominance: A review of clinical and experimental evidence. In J. Gruzelier & P. Flor-Henry (Eds.), *Hemispheric asymmetries of function in psychopathology.* Amsterdam: Elsevier.

Goffman, E. (1963). *Stigma: Notes on the management of spoiled identity.* Englewood Cliffs, N.J.: Prentice-Hall.

Goldenson, R. M. (1978). *Disability and rehabilitation handbook.* New York: McGraw-Hill.

Graham, P. (1985). Handling stress in the handicapped adolescent. *Developmental Medicine and Child Neurology, 27* (3), 389–391.

Green, M. & Solnit, A. J. (1964). Reaction to the threatened loss of a child: A vulnerable child syndrome. *Pediatrics, 34,* 58.

Guttman, L. (1976). *Textbook of sports for the disabled.* London: H. M. & M. Publishers.

Heilman, K. M. & Valenstein, E. (1984). *Clinical neuropsychology.* New York: Oxford University Press.

Hewett, S. (1970). *The family and the handicapped child.* London: Allen & Unwin.

Ispa, J. (1981). Social interaction

among teachers, handicapped children and non-handicapped children in a mainstream preschool. *Journal of Applied Development in Psychology, 1*, 231–250.

Jessell, J. C. & Veltri, F. J. (1986). Trait anxiety among physically disabled adolescents. *Journal of Rehabilitation, 6*, 45–49.

Katz, A. M. (1961). *Parents of the handicapped child.* Springfield, Ill.: C. C. Thomas.

Kessler, J. (1977). Parenting the handicapped child. *Pediatric Annals, 6*, 10.

Klapper, Z. S. & Birch, H. G. (1966). The relation of childhood characteristics to outcome in young adults with cerebral palsy. *Developmental Medicine and Child Neurology, 4*, 643.

Klaus, M. H. & Kennel, J. H. (1976). *Maternal-infant bonding.* St. Louis: C.V. Mosby.

Kogan, K. L., Tyler, N. & Turner, P. (1974). The process of interpersonal adaptation between mothers and their cerebral palsy children. *Developmental Medicine and Child Neurology, 16*, 518.

Kolin, T., Scherzer, A., New, B. & Garfield, M. (1971). Studies of the school age child with meningomyeloceles, social and emotional adaptation. *Journal of Pediatrics, 78*, 1013.

Korsch, B. M. & Negrete, V. F. (1972). Doctor-patient communication. *Scientific American, 227*, 66.

Lemkau, P. (1961). The influence of handicapping conditions on child development. *Children, 8*, 43–47.

McDaniel, J. W. (1976). *Physical disability and human behavior.* Elmsford, N.Y.: Pergamon.

McDonald, E. T. (1962). *Understand those feelings.* Pittsburgh: Stanwix House.

McKeith, R. (1973). The feelings and behavior of parents of handicapped children. *Developmental Medicine and Child Neurology, 15*, 524.

McKeith, R. (1976). The restoration of the parents as the keystone of the therapeutic arch. *Developmental Medicine and Child Neurology, 18*, 285.

McMichael, J. K. (1971). *Handicap: A study of physically handicapped children and their families.* Pittsburgh: University of Pittsburgh Press.

Magill, J. & Hurlbut, N. (1986). The self-esteem of adolescents with cerebral palsy. *American Journal of Occupational Therapy, 40*, 402–407.

Mahler, M. L. (1963). Thoughts about development and individuation. *Psychoanalytic Study of the Child, 18*, 307.

Meadow, K. P. & Meadow, L. (1971). Changing role perceptions for parents of handicapped children. *Exceptional Children, 38*, 21.

Milavic, G. (1985). Do chronically ill and handicapped children become depressed? *Developmental Medicine and Child Neurology, 27*(5), 677–682.

Milunski, A. (Ed.). (1981). *Coping with crisis and handicap.* New York: Plenum.

Minde, K. K. (1978). Coping styles of 34 adolescents with cerebral palsy. *American Journal of Psychiatry, 135*, 1340–1344.

Minde, K. K., Hackett, J. D., Killou, D. & Silver, S. (1972). How they grow up: 41 physically handicapped children and their families. *American Journal of Psychiatry, 128*, 1554–1559.

Molnar, G. E. (1979). Cerebral palsy: Prognosis and how to judge it. *Pediatric Annals, 10*, 40–50.

Molnar, G. E. (1981). Intervention

for physically handicapped children. In M. Lewis & L. T. Taft (Eds.), *Developmental disabilities: Theory, assessment and intervention.* New York: Spectrum Publications.

Molnar, G. E. (1985a). Rehabilitative benefits of sports for the handicapped. *Connecticut Medicine, 49,* 574–578.

Molnar, G. E. (1985b). Cerebral palsy. In G. E. Molnar (Ed.), *Pediatric rehabilitation.* Baltimore: Williams & Wilkins.

Mueller, J. (1983). Neuroanatomic correlates of emotion. In L. Temoshok, C. Van Dyke & L. S. Zegans (Eds.), *Emotions in health and illness.* New York: Grune & Stratton.

Ortigas, M. T. (1983). The physical and emotional problems of persons afflicted with cerebral palsy. *Philippine Journal of Education, 62,* 322–328.

Osman, B. & Blinder, H. (1982). *No one to play with: The social side of learning disabilities.* New York: Random House.

Oswin, M. (1967). *Behaviour problems amongst children with cerebral palsy.* Bristol: Wright & Sons.

Parker, B. (1971). A case of congenital spina bifida. Impact of the defect on psychic development. *International Journal of Psycho-Analysis, 52,* 307.

Pearlman, L. & Scott, K. A. (1981). *Raising the handicapped child.* Englewood Cliffs, N.J.: Prentice-Hall.

Perlman, J. L. & Routh, D. K. (1980). Stigmatizing effects of a child's wheelchair in successive and simultaneous interactions. *Journal of Pediatrics, 5,* 43–55.

Perrin, J.C.S., Reusch, E. L., Pray, J. L., Wright, G. F. & Bartlett, G. S. (1972). Evaluation of a ten-year experience in a comprehensive care program for handicapped children. *Pediatrics, 50,* 793.

Reynell, J. (1973). Children with physical handicaps. In V. P. Varma (Ed.), *Stresses in children.* New York: Crane, Russak.

Richardson, S. A. (1969). The effect of physical disability on the socialization of a child. In D. A. Goslin (Ed.), *Handbook of socialization: Theory and research.* New York: Rand McNally.

Richardson, S. A. (1970). Age and sex differences in values towards physical handicaps. *Journal of Health and Social Behavior, 11,* 207–214.

Richardson, S. A. (1971). Research report: Handicap, appearance and stigma. *Social Science and Medicine, 5,* 621–628.

Richardson, S. A. (1972). People with cerebral palsy talk for themselves. *Developmental Medicine and Child Neurology, 14,* 524.

Richardson, S. A., Goodman, N., Hastorf, A. M. & Dornbusch, S. M. (1961). Cultural uniformity in reaction to physical disability. *American Sociological Review, 26,* 241.

Richardson, S. A., Ronald, L. & Kleck, R. E. (1974). The social status of handicapped and non-handicapped boys in a camp setting. *Journal of Special Education, 8,* 143–152.

Robinault, I. P. (1978). *Sex, society and the disabled.* New York: Harper & Row.

Rubinstein, B. (1982). Psychiatric aspects of physical handicaps. In J. A. Downey & N. L. Low (Eds.), *The child with disabling illness. Principles of rehabilitation.* New York: Raven.

Sears, R., Maccoby, K. E. & Levin, H. (1957). *Patterns of child rearing.* Evanston, Ill.: Row, Peterson.

Seidel, V., Chadwick, O. & Rutter, M. (1979). Psychological disorders in crippled children. In S. Chess & A. Thomas (Eds.), *Annual progress in child psychiatry and child development*. New York: Brunner/Mazel.

Shakespeare, R. (1975). *The psychology of handicap*. London: Methuen.

Shere, E. S. (1971). Patterns of child rearing in cerebral palsy, effects upon the child's cognitive development. *Pediatric Digest, 23*, 28.

Shere, E. S. & Kastenbaum, R. (1966). Mother-child interaction in cerebral palsy, environmental psychosocial obstacles to cognitive development. *Genetic Psychology Monographs, 73*, 255.

Skinner, B. F. (1974). *About behaviorism*. New York: Vintage Books.

Solnit, A. J. & Stark, M. H. (1961). Mourning and the birth of a defective child. *Psychoanalytic Study of the Child, 15*, 523-557.

Springer, S. P. & Deutsch, G. (1981). *Left brain, right brain*. San Francisco: W. H. Freeman.

Strain, P. S. & Kerr, M. M. (1979). Treatment issues in the remediation of handicapped preschool children's social isolation. *Education and Treatment of Children, 2*, 197.

Sussman, M. B. (1965). *Sociology and disability*. Washington, D.C.: American Sociological Association.

Swisher, L. & Pinscher, E. (1981). The language characteristics of hyperverbal hydrocephalic children. *Developmental Medicine and Child Neurology, 13*, 746.

Taft, L. T., Matthews, S. W. & Molnar, G. E. (1983). Pediatric management of the physically handicapped child. In L. A. Barnes (Eds.), *Advances in pediatrics* (Vol. 30). Chicago: Yearbook Publishers.

Teplin, S. W., Howard, J. A. & O'Conner, M. (1981). Self-concept of young children with cerebral palsy. *Developmental Medicine and Child Neurology, 23*, 730-738.

Tew, B. J. & Laurence, K. M. (1973). Mothers, brothers and sisters of patients with spina bifida. *Developmental Medicine and Child Neurology (Supplement), 29*, 69.

Tew, B. J. & Laurence, K. M. (1976). The effect of admission to hospital and surgery on children with spina bifida. *Developmental Medicine and Child Neurology (Supplement), 37*, 119.

Tew, B. J., Payne, H. & Laurence, K. M. (1974). Must a family with a handicapped child be a handicapped family. *Developmental Medicine and Child Neurology (Supplement), 32*, 95.

Vanderheiden, G. C. & Grilley, K. (1983). *Non-vocal communication techniques and aids for the severely physically handicapped*. Baltimore: University Park Press.

Van Putte, A. W. (1979). Relationship of school setting to self-concept in physically disabled children. *Journal of School Health, 49*, 576-578.

Wasserman, G., Allen R. & Solomon, C. R. (1985). At-risk toddlers and their mothers: The special case of physical handicap. *Child Development, 56*, 73-83.

Wasserman, G. A. (1986). Affective expression in normal and physically handicapped infants. Situational and developmental effects. *Journal of the American Academy of Child and Adolescent Psychiatry, 25*(3), 393-399.

Wolraich, M. L. (1982). Communication between physicians and parents of handicapped children. *Exceptional Children, 48,* 316.

Wortis, H. Z. & Cooper, W. (1957). The life experience of persons with cerebral palsy. *American Journal of Physical Medicine, 36,* 28.

# 6
# Psychosocial Issues in the Treatment of Children with Muscular Dystrophy

*Gloria D. Eng*

Duchenne muscular dystrophy (DMD) is a progressive disability of early onset, which remains one of the most difficult challenges in medicine. It is a disease that demands not only anticipatory medical guidance but also sustained compassion and psychosocial support for the affected individual and for the entire family—over an extended period. This chapter will focus on the clinical course, diagnosis, and therapeutic management of a child with Duchenne muscular dystrophy—with particular emphasis on the psychosocial problems that are pervasive throughout the illness.

Duchenne (1868) described his first case of muscular dystrophy as "paraplegie hypertrophique de l'enfance de cause cerebrale." He presumed a cerebral origin for the disorder because of the associated intellectual retardation in the child. In a subsequent detailed study of 13 cases, he defined it as a condition more frequently affecting male children, who show (1) progressive weakness of movements, initially affecting the muscles of the lower limbs and lumbar spine, gradually getting worse and spreading to the upper limbs; (2) enlargement of some of the paralyzed muscles (pseudohypertrophy); and (3) hyperplasia of the interstitial connective tissue in the paralyzed muscles with scarring and adipose tissue deposition in the late stages. Gowers (1879) lectured on the subject, describing it as "one of the most interesting, and at the same time most sad of all those with which we have to deal."

## INCIDENCE AND GENETICS

The exact incidence of Duchenne muscular dystrophy in the United States is unknown. The Muscular Dystrophy Association (D. S. Wood, pers. com. 1985) estimates that 12,000 to 20,000 children have the disorder in this country, suggesting an incidence of 1 in 4,000 live male births. Recent investigations by Monaco et al. (1986) and Davies et al. (1987) have located the gene for Duchenne and Becker muscular dystrophy on the short arm of the X chromosome at X p 21. Further highly significant studies by Hoffman et al. (1988) have identified in Duchenne muscular dystrophy the missing protein, subsequently labeled dystrophin. These findings have allowed more precise identification of carriers of the dystrophies, the differentiation of the various types, and the determination of whether a fetus in utero is affected. Genetic counseling is an integral part of medical management in the family who has a child with DMD.

## CLINICAL MANIFESTATIONS

The clinical course (table 6.1) of the child with DMD is one of progressive weakness. A rare child will present with delayed onset of walking (beyond 18 months of age). Most are suspect for abnormality between 2 and 4 years of age because of an abnormal gait, frequent falls while attempting to walk or run, and difficulties ascending or descending stairs. The affected child develops increasing hip girdle weakness, which manifests by a wider-based stance, waddling gait, and hyperlordosis of the lumbar spine. He gradually thrusts up on his toes to accommodate the anterior shift of his center of gravity. In a further compensatory attempt to maintain balance, he throws his shoulders backward. Some children are first brought to the physician's attention because of a speech impediment or because of an incapacity to learn in school. The hypertrophy of muscles that include the shoulder girdle and calves may affect the tongue as well. The learning disability is related to retardation: approximately one third of affected boys are retarded, some severely. The group's mean IQ on the Wechsler scale is 85, normal being 100 (Dubowitz, 1965).

Between 5 to 7 years of age there may be a hiatus when the deterioration is less apparent. The process is variable in that some children become weaker more rapidly than others. They finally lose the capacity to walk between 8 and 12 years of age. Once relegated to a wheelchair-bound

**TABLE 6.1**
Clinical Course of Duchenne Muscular Dystrophy

| Age | Condition |
|---|---|
| 2–4 | Abnormal gait<br>Frequent falls<br>Difficulty with stairs |
| 5–7 | Hiatus |
| 8–12 | Increasing hip girdle weakness (waddle, hyperlordosis)<br>Pseudohypertrophy<br>Shoulder girdle weakness |
| Adolescence | Pulmonary and cardiac problems<br>Scoliosis |
| 20–30 | Demise |

existence, the weakness of these children inexorably increases. They then develop restrictive lung disease, and the dystrophic process affects the muscles of the heart as well. Soft-tissue contractures and bony misalignment such as scoliosis of the spine become serious management problems. The majority of affected children die by 20 years, a few live beyond their 20s, and a few even to 30 years of age as new technological advances allow artificial ventilation to be a practical option (Bach et al., 1987). Others die suddenly at a younger age secondary to myocardial insufficiency or respiratory failure usually associated with an intercurrent pulmonary infection. There is also a congenital form of muscular dystrophy, which affects infants and young children and leads to a severely curtailed life span.

## DIAGNOSIS

Clinical investigations, which are necessary to confirm the diagnosis of DMD, comprise a detailed history with emphasis on family history, a physical examination including testing of muscle strength and functional capabilities, a study of serum enzyme levels, electrodiagnosis, and muscle biopsy. In DMD there is extreme elevation of several serum enzymes including the aminotransferases, aldolase, lactate dehydrogenase, and creatinekinase. The last is particularly elevated in the early stages of the disease, its levels declining as muscle tissue is lost with progression of the disease. Electromyography shows increased recruitment of motor unit action potentials for effort in contraction; low-amplitude and short-duration potentials; and increase in polyphasic potentials. Occasionally fibrillation potentials and satellite potentials may be seen. The muscle biopsy confirms the diagnosis and usually demonstrates variability in fiber size, evidence of regeneration or degeneration of fibers, presence of internal nuclei, fiber splitting, increase in interstitial connective tissue, and lipocytes.

## THERAPEUTIC MANAGEMENT

The basic principle of therapeutic management of the child with a progressive disability such as DMD includes a sensitivity or awareness of the psychological issues affecting the child and his family over time. The emphasis in rehabilitation is on function despite disability; the attitude should be positive, anticipatory, and always supportive.

At the early encounter with the child and his family, the diagnosis may be suggested but should not be confirmed until the results of enzyme studies, electrodiagnosis, and biopsy are completed. Both parents should be present at the confirmatory session; it is unfair to expect the mother or father to be alone when confronted with this diagnosis. Other adult members of the family may have to be included, especially if there is a single parent. The reactions to diagnosis include shock, disbelief, denial, hysteria, rage, and indescribable anguish or grief. Occasionally, the reaction is one of stony silence. It is sometimes helpful to leave the parents alone after the initial disclosure of the diagnosis to allow time and privacy for them to collect themselves. They may be so stunned that there is no immediate comprehension of the medical explanations. Some parents have stated, "I never heard anything you told me!" Serial visits are usually necessary to answer questions and direct the parents to appropriate sources for second opinions if they so request.

A physical therapeutic exercise program with emphasis on appropriate recreational activities is initiated. The child may sense the concern and anxiety in his parents and react by becoming more dependent and immature and

## TABLE 6.2
Therapeutic Management of Children with Duchenne Muscular Dystrophy: Medical, Social and Psychological Aspects

| Age | Reactions of Child | Reactions of Parents | Strategies |
|---|---|---|---|
| 2–4 | Immaturity, dependency, frustration | Shock, disbelief, unacceptance, hysteria, anger, grief, anguish | Allow time and space; arrange serial visits; help with second opinion; initiate physical therapy; instruct on appropriate recreation |
| 5–7 (Hiatus) | Slightly increasing weakness; anger; acting out frustration; awareness of something wrong | Mother's guilt; father's problems with acceptance; involvement with parent groups, muscular dystrophy agencies; sublimation vs. social isolation | Reinforce participation in school, physical therapy, sports, camp; provide adaptive seating and other assistive devices; join parent groups; use social services |
| 7–12 | Self-image problem, shame, depression, loss of control, increased dependency, acting out, verbal abuse | Parents coping in different ways; father taking extra jobs; mother struggling with daily life and special problems; siblings frightened and resentful | Attend to diet, surgery, bracing vs. adaptive seating; address spinal problems; emphasize hobbies, outings; involve child in his own care; be truthful, predictable; offer no false promises; deal with pulmonary and cardiac function |
| 12–20 (crisis period) | Struggling for independence in the face of increasing dependency; loss of body; passivity; isolation; denial; tenacity for life | Inevitability of progression; exhaustion; franticness; anger at caregivers; hopelessness; depression; hidden strength | Physical therapy: transfers, breathing exercises, postural drainage as needed, mobility of limbs, skin care; activities of daily living from chair; mobility in chair in sports; group psychotherapy; group independent living arrangements; vocational and school evaluation |
| Terminal | Struggling for survival vs. resignation | Loss, grief, relief | ICU problems with or without ventilator; availability of support system for entire family |
| Follow-up | | | Check on entire family; counsel carrier daughters |

by manifesting his frustrations in frequent temper tantrums. It is advisable to assign the family to a social worker to help with the immediate and future problems in familial relationships.

Between 5 and 7 years of age there often is a hiatus wherein the child's condition appears to be fairly static. He enters school and realizes that he is different from the other children and cannot participate in many of the usual sports activities of his peers. His awareness that something is wrong within himself is heightened when he stumbles and falls a great deal and becomes the butt of his classmates' jokes. He can react by assuming the role of the class clown or becoming hostile, angry, and depressed. The parents meanwhile are struggling with the daily exigencies of raising a family as well as tending to this child's special problems. The catastrophic impact of the diagnosis of DMD has lessened. The mother learns to deal with the guilt imposed on her by the fact that she is the genetic carrier of the disease. She may already have experienced the disorder in a brother or uncle. She may or may not have a supportive and nonaccusatory husband. The father of the child has to deal with disappointment and ego deflation when confronted with his affected son. To compensate, he will sometimes take several jobs with the excuse of raising enough money to pay for the medical expenses, thereby relegating the daily care of the child to the wife. Not infrequently, however, parents can sublimate their loss and pervasive grief by becoming advocates for their child and others like him. They can become involved in parents groups, in the Muscular Dystrophy Association (MDA), or as leaders in recreational activities and adaptive sports programs. They may join legislative groups to lobby for the rights of the disabled. Other parents become more reclusive, isolating themselves from friends and relatives in their effort to accommodate their perceived shame and misfortune.

The child, meanwhile, endures the routine of daily physical therapy, which is given to maintain the range of motion of his joints; to stretch various muscles such as the iliopsoas, tensor fasciae latae, Achilles tendons, and finger flexors; and to provide frequent encouragement to keep good posture and alignment. The parents are taught to perform postural drainage, including breathing exercises, to help the child clear his lungs in the event of respiratory infections. Then comes the need to adapt seats, replace steps with ramps, adjust the height of toilets and sinks, design clothing to maintain independence in dressing, and customize other activities of daily living. The school must be made aware of the child's condition, and special educational techniques may have to be applied to the child who is retarded or has learning disabilities. Certain sports activities, especially swimming and summer camp programs designed for the child with DMD, are recommended.

Between 7 and 12 years of age, the child's ability to walk progressively deteriorates. Physicians need to elicit the child's compliance and the parents' in the selection of appropriate bracing to maintain erect ambulation. Surgical releases of tendons may precede the brace fitting. The decision to maintain ambulation at all cost, must be weighed against forfeiture of ambulation for greater mobility in a standard or motorized wheelchair. The boy is counseled that he must maintain as much movement as possible even from a chair, lest he lose the strength he does not use. Overhanging trapeze bars and side rails on the beds are necessary to help him shift position when lying down. Before too long, his parents will have to turn him during the night. Still, active elicitation of the boy's cooperation in making each change is important to allow for a semblance of autonomy. Medical concerns may focus on dietary restrictions as immobility leads to increasing obesity in some children. The physician must deal with the restrictive lung disease that appears as the respiratory muscles weaken; the cardiomyopathy and incipient heart failure; the increasing muscle contractures, which can lead to subluxated or dislocated hips; and the scoliosis of the spine, which is almost inevitable in the child who spends his life in a wheelchair.

The years between 12 and 20 are punctuated by periods of crisis. The struggle for independence in the face of increasing dependency continues. The child experiences a strong sense of "loss of body." The need for privacy is thwarted by the necessity of the parents, particularly the mother, to give the most intimate and personal care such as bathing, toileting, dressing, and lifting from bed to chair and chair to bed. The young man becomes more passive and isolated. For the parents, progression of the disease becomes more obvious. Exhaustion ensues from providing unremitting care. Their hopelessness and depression may be expressed in retaliatory anger and frustration at the physician and other medical personnel.

At this time, the medical effort nevertheless has to be reinforced. The physical therapist helps to teach the parents and the child safe transfer activities, with modification of living arrangements to ensure the privacy of the child. There is continued encouragement for independence and self-help activities. Maintenance of range of motion of the joints, prevention of skin breakdown, and attention to posture and spinal alignment continue. Spinal support braces, special attachments to the wheelchair, the selection of the wheelchair itself, and the choice of a proper bed remain cooperative ventures between the child, his parents, and medical personnel. School performance and vocational capabilities are evaluated. As the boy grows into adulthood and should his health remain relatively stable despite weak musculature, he frequently will desire to go away to school or live in an independent living arrangement. Special housing facilities are becoming more available for the disabled, and parents are encouraged to allow for these possibilities. Adaptive wheelchair sports, recreational outings, involvement in peer activities, and group psychotherapy are fostered. Because DMD is not a painful disorder, it is easier for the affected person to deny that something is really wrong. Tenacity for survival remains, and the desire for a life-style that approximates normal persists.

Directed group psychotherapy sessions have proven of great value to the disabled adolescent with DMD. He frequently sees his disabled peers as abnormal and himself as intact. As he learns to accept the others as persons not unlike himself, he becomes more able to deal with his own incapacity. The sessions allow for the ventilation of frustration, anger, disgust, problems of sexuality, and problems of excessive control by parents, siblings, and physicians. Some of the members have great resiliency and strength and are able to share with and support the more emotionally tenuous members of the group. Humor remains the ultimate sustaining force. For the rest of the family, parent and sibling support groups can go on simultaneously.

The terminal phase is highly individual for each young person with DMD and his family. Resignation and acceptance of the inevitable are not the rule especially in this day of high technology and options for extending life through mechanical support systems.

*Case 1.* S.Y., a 17-year-old, mildly retarded black male was the second affected child in the family. His brother had died of DMD 5 years previously. He and his mother had developed a very special relationship—he never resented her attentiveness, and she enjoyed his open and spontaneously affectionate demeanor. She was not always conscientious about the medical aspects of his care, but she provided him full emotional, social, and recreational support while maintaining full-time employment. She included him in all special family excursions and functions. When he became increasingly weak and cachectic, requiring repeated admissions to the hospital, he was terrified of death. In the end, he knew and stated that he was ready to join his brother. Family members, including friends and clergy, gathered around his bedside; his mother held him and soothed him until he breathed his last. It was a happening. Though the mother was filled with a sense of loss and sorrow, there was an elation and a relief that his brief life—so tragically burdened—was over.

*Case 2.* S.B. is a 20-year-old boy, intelligent, motivated, and seriously afflicted with the end stages of DMD. Because of years of haphazard medical attention, he has serious contractures, progressive kyphoscoliosis, restrictive lung disease, and cardiac decompensation. On his latest admission to the hospital, he chose tracheostomy and ventilator support. He was able to go home with full nursing support, nasogastric feedings, suctioning, and artificial ventilation. He has managed to control his own environment with aplomb, to maintain artistic pursuits, and to educate his family members in his care, and he is surrounded by the comings and goings of innumerable friends. He maintains that each day he survives, he is closer to the possibility of a breakthrough in medical science that will cure his disease.

The concern of physician care givers for the family who loses a son with DMD is not over at the time of death. The mother and not infrequently the father, who for years have worked very hard in providing constant care, may never have had time during the child's lifetime to properly grieve. Suddenly the task is over; the void is frequently profound and the grief unassuaged. Physicians need to remain available to the parents and other members of the family for months and even years following the demise of the affected child.

**REFERENCES**

Bach, J. R., O'Brien, J., Krotenberg, R. & Alba, A. S. (1987). Management of end-stage respiratory failure in Duchenne muscular dystrophy. *Muscle and Nerve, 10,* 177–182.

Davies, K. E., Farrest, S., Smith, T., Kenwrick, S., Ball, S., Dorkins, B. M. & Patterson, M. (1987). Molecular analysis of human muscular dystrophy. *Muscle and Nerve, 10,* 191–199.

Dubowitz, V. (1965). Intellectual impairment in muscular dystrophy. *Archives of Disabled Children, 40,* 296–301.

Duchenne, G. B. (1868). Recherches sur la paralysie musculaire pseudo-hypertrophique ou paralysie myosclérosique. *Archives Générales de Médicine, 11,* 5–25, 179–209, 305–321, 421–443, 552–588.

Gowers, W. R. (1879). Clinical lecture on pseudo-hypertrophic muscular paralysis. *Lancet, 2,* 1–2, 37–39, 73–75, 113–116.

Hoffman, E. P., Fischback, K. H., Brown, R. H., Johnson, M., Medori, R., Loike, J. D., Harris, J. B., Waterston, R., Brooke, M., Specht, L., Kupsky, W., Chamberlain, J., Caskey, T., Shapiro, F. & Kunkel, L. M. (1988). Characterization of dystrophin in muscle biopsy specimens from patients with Duchenne or Becker muscular dystrophy. *New England Journal of Medicine, 318*(21), 1363–1368.

Monaco, A. P., Neve, R. L., Colletti-Feener, C., Bertelson, C. J., Kurnit, D. M. & Kunkel, L. M. (1986). Isolation of candidate cDNAs for portions of the Duchenne muscular dystrophy gene. *Nature, 323,* 646–650.

# 7

# Psychosocial Adjustment to Spinal Cord Injury

*Roberta B. Trieschmann*

Spinal cord injury is a low-incidence but high-cost disability that imposes tremendous changes on a person's lifestyle. The majority of persons who face these changes are young, active males. Young et al. (1982) report that males account for 82% of the spinal injuries in the United States. A majority of these injuries (62.5%) are acquired by persons between 15 and 29 years. Within this age group the etiologies are approximately: vehicular/pedestrian accidents (52%), penetrating wounds (14%), sports-diving (13%), sports-other (6%), falls (11%), falling or flying objects (4%), and other causes (2%). The incidence of paraplegia is 47% and quadriplegia, 53%. Paraplegia and quadriplegia occur at equal rates for most of these accidents except for four categories. Penetrating wounds are more likely to result in paraplegia (72%) than quadriplegia, as are accidents associated with falls or flying objects (72% paraplegia). However, sports injuries are to a dramatic degree more likely to result in quadriplegia. Ninety-eight percent of the diving injuries and 80% of other sports injuries result in quadriplegia. Although women account for a small percentage of the spinal injuries, it is interesting to note that their etiologies are somewhat different from those of their male counterparts: vehicular/pedestrian accidents (60%), penetrating wounds (14%), falls (16%), and other causes (10%). These data were gathered from 1973 to 1981 at the Model Regional Spinal Injury Treatment Systems in the United States and thus may be slightly biased in favor of incidence of quadriplegia since the more difficult and complicated injuries were more likely to be referred to such specialized treatment centers. However, the general distribution of the data should be fairly representative of the population seen in most major hospitals and rehabilitation centers (J. Young, pers. com. Feb. 7, 1984).

Thus, the affected population is young and predominantly male, and the injuries occur through vigorous activities in a large number of cases, with misjudgment being a factor in many of them. It should be noted, however, that there are no data to suggest a tendency toward a psychological self-destructiveness in this group. Rather, there is a high base rate of injudicious motor behavior during the teenage years and the early 20s in males, and a subset of these individuals acquire a spinal cord injury. This should not be confused with self-selection. Considering the advances in medical science, this group can expect to live almost a normal life span, but the quality of that life will be influenced by a variety of psychological, social, and environmental variables (Trieschmann, 1987, 1988).

The persons who acquire a spinal cord injury are often in the midst of mapping out a career or course of action that will characterize their adult lives, when they suddenly find themselves paralyzed, with no sensation in their limbs and no control over bladder and bowel. Life as they had known it will be interrupted by months of hospitalization and an often lengthy period during which new techniques must be mastered for survival and independent function. The changes in life-style will be significant. Spinal-injured persons must learn how to deal with a world designed for and dominated by able-bodied persons who are not very accepting of those with disabilities. They must learn to face acquaintances who now suggest that they are different from how they used to be and perhaps "less". They must learn new types of recreation and leisure activities and, in many cases, a new vocation. However, in following an educational or vocational training program they must learn to face potential employers who do not want to hire them, not because of lack of qualifications for the job but because of the disability. They must learn that many of their previous friends drift away, and thus they must seek opportunities to meet new people and to make new friends. Yet strangers tend to avoid any interaction with them; consequently, new techniques must be learned to put others at ease and to make them forget the presence of the wheelchair. Disabled persons must learn a sense of humor in order to cope with the daily frustrations and hard work that living with a disability entails. And, they must learn to maintain their sense of dignity and self-worth when faced with a social welfare system that penalizes their efforts to become independent and self-sufficient.

This, then, constitutes the impact of a disability on a person's life, and there are many factors that influence the ultimate adjustment the person will make. In this chapter, we will consider some of the psychosocial variables that influence adjustment. We must note that it would be a mistake for health care professionals to assume that these are the only issues that influence the outcome of rehabilitation efforts (Trieschmann, 1988; Trieschmann & Willems, 1980).

## THE REHABILITATION PROCESS

Rehabilitation is the process of learning to live with one's disability in one's environment. This learning experience is a dynamic process that starts at the moment of injury and continues for the remainder of the person's life. There is no definable end point that allows the person to be labeled "rehabilitated" or "adjusted" because, as with all people, disabled persons continually need to adapt to their environment in more functional and satisfying ways. It may take as long as 2 years for the new activities of daily living and mobility techniques to become an automatic part of daily life, and this is without considering the changes in social, recreational, and vocational behaviors that the person must incorporate into his or her repertoire of functional activities. Thus, it is most important for health care professionals to understand the principles of learning and the multiplicity of factors that are involved in the process of adjusting to a disability. In the first several weeks after the onset of spinal cord injury, survival of the patient is the concern of hospital personnel, and the person becomes the passive recipient of treatment designed to fix his or her body (e.g., skeletal traction or surgery, treatment of associated injuries, management of bladder and bowel, prevention of skin problems). When medical stability has been achieved and the newly disabled person is no longer sick, a rehabilitation program to teach the person how to manage the activities of daily living and the mobility techniques to negotiate the world will be outlined. At this point the person can no longer be the passive recipient of dispensed treatments but must become an active participant in the process of learning to live with the spinal injury.

Unfortunately, the operational policies and procedures of hospitals and rehabilitation centers have been designed to dispense units of organic treatment. The person is a patient and the staff deliver treatments, usually according to a schedule and sequence designed by the staff for their convenience. This is not an optimal environment in which to teach new behaviors to persons who are not sick but who essentially are students. Therefore, two models of rehabilitation will be examined so that we can determine which provides the better fit between the task at hand and the strategies used to accomplish that task.

The *medical model* of rehabilitation could be stated as:

$$B = f(O \times p)$$

Behavior (B), or adjustment, of the spinal-injured person is the result of treatments dispensed to the organic (O) variables (skin, bladder and bowel, paralysis, lack of sensation, respiratory function, etc.) unless hindered by underlying personality problems (lack of motivation, depression, low self-esteem, anxiety, anger, frustration, dependency). In this case, units of treatment will be dispensed to the personality (p) variables in the form of counseling and psychotherapy.

In contrast to the medical model is the *educational model* of rehabilitation, the learning approach. In this instance, rehabilitation is viewed as the process of teaching the person to live with the disability in his or her own environment. The person must be an active participant in this process, and the program must be designed by the staff with, not for, the person to meet his or her needs and goals. These will be determined by an assessment of the person's unique personality style, desires, and preferences and the environment to which he or she will return. If rehabilitation is the process of teaching a person to live with a disability, then the principles of learning and the multiple factors that influence behavior become the concern of everyone on the rehabilitation team, not just the psychologist. The learning or educational model of rehabilitation can be summarized as:

$$B = f(P \times O \times E)$$

Behavior (adjustment) is a function of the interaction of person variables (P), organic variables (O), and environmental variables (E). Person variables would include: habits, locus of control (defined later), method of coping with stress, preferences and rewards, self-image, creativity. Organic variables would include: level of injury, age, medical complications, strength and endurance. Environmental variables would include: hospital milieu; stigma value of the disability; family and interpersonal support; financial security; cultural and ethnic influences; access to medical attention, equipment repair, and recreational and educational opportunities; architectural barriers; transportation.

Traditionally we have focused on treating the organic problems and assumed that all responsibility for success or failure rested with the person.

This is the essence of the medical model. Perhaps it is most appropriate to use the medical model during the acute phase of treating illness and disability. However, this model is not appropriate during the rehabilitation phase during which a variety of behaviors (B) must be learned to enable the person to resume a satisfying life as an integrated part of his or her community. Using the learning model, $B = f(P \times O \times E)$, we need to pay attention to the multiple factors that influence behavior, particularly the environmental variables, which have a powerful impact on rehabilitation. Thus, within this context, the psychological variables are only one part of the equation that influences a person's behavior. An understanding of the role of psychosocial influences on the outcome of rehabilitation efforts is important. The most important issues will be reviewed in this chapter.

## PSYCHOSOCIAL REACTIONS TO THE ACUTE INJURY

There has been considerable speculation but little research to document the immediate reaction to spinal cord injury. Theorists have proposed a stage theory of adjustment to disability, with denial of the physical implications of the injury being the first stage (Gunther, 1969; Hohmann, 1975; Kerr & Thompson, 1974; Siller, 1969). A significant percentage of persons with new injuries ask questions or make statements that suggest they do not perceive themselves to be paralyzed or the situation to be permanent. They talk about walking out of the hospital or ask repeatedly, "When will I be able to move my legs?" However, the question needs to be raised whether this constitutes evidence of the psychological defense mechanism of denial or whether there are other explanations for this behavior.

The immediate reaction to spinal cord injury should be examined in terms of the psychological reaction to not only the injury itself but the procedures used to ensure the survival of the person during the acute treatment phase. For example, paralysis restricts movement, and the procedures used to immobilize the spine will further restrict movement for at least 8 weeks. Persons with quadriplegia who have tongs in their head to immobilize the neck have less movement possible than persons with paraplegia, and furthermore, they have a more restricted visual field: they can look at the ceiling or floor only. Medications to relax the person, to ease the pain, and to treat associated injuries and medical problems may cloud the sensorium for weeks after the accident. Later in the course of disability it is not unusual for high dosages of valium to be given to control muscle spasms; this can further reduce mental acuity. Anesthesia for surgery followed by treatment in intensive care units can further deteriorate mental functioning.

Intensive care units provide little opportunity to rest comfortably because of the frequent interruptions required for medical procedures. In fact, for many months the person with spinal injury will be awakened every 2 hours to be turned, in order to prevent skin lesions. Continued loss of sleep will disrupt mental efficiency further. Since the patient has few cues to identify the passage of time (no windows and continuous artificial lighting), disorientation can occur.

Loss of sensation associated with the injury is one form of sensory deprivation; the restricted view of the world (ceiling or floor) while in traction is another. The monotony of the hospital routine is certainly a perceptual restriction as is the lack of intellectual challenge that accompanies all long hospitalizations. Consequently, research on the effects of sensory deprivation is relevant to understanding the behavior of the newly injured person (Zubek, 1969). Zubek's findings indicate that restriction of movement does produce stress, particularly bodily discomforts and thinking difficulties, but that sensory deprivation is associated with significantly more stress. Physical exercise when compared with no exercise during periods of perceptual deprivation is associated with significantly less impairment in intellectual and perceptual motor tests and fewer EEG changes. Furthermore, 2 weeks of perceptual deprivation is associated with a progressive decrease in mean occipital lobe frequencies on the EEG. The EEG frequencies begin to increase when perceptual stimulation increases, but even after 10 days they may remain below normal. Correlated with the magnitude of the EEG changes are motivational losses such as an inability to concentrate on or engage in purposeful activity (Zubek, 1969).

Newly injured persons have frequently been described as denying the disability, because they ask what has happened to them, receive an explanation, and act as if they had never received an answer. However, rather than attribute this to deep psychological processes, it seems more parsimonious to interpret this "forgetfulness" as a side effect of sensory deprivation during the acute treatment period.

Harris et al. (1973) confirm these observations regarding the acute treatment phase and add that pain is a frequent concomitant of spinal injury, which interferes with focused thinking. In addition, fear of dying is a realistic concern in many cases, and respiratory complications plus associated injuries may bring survival into question at several points in the immediate recovery phase.

In this context, then, it is interesting to note that Braakman et al. (1976) report that most newly injured persons that they interviewed wanted information on their condition and on future implications within a few weeks of their spinal injury. There was no evidence of denial of the disability, and once the intellectually obtunding effects of the early treatment procedures had dissipated, the individuals wanted to know just what had happened to them.

Nevertheless, even after receiving a complete explanation of the spinal injury, many persons adamantly claim they will walk out of the hospital. Is this denial or hope? Perhaps we have a semantic problem. Let us define denial as a nonrecognition of the implications of the injury to such an extent that the person perceives little need to participate in rehabilitation. In other words, "I am not permanently paralyzed and will not need these rehabilitation therapies, because I am going to recover completely." In the author's experience, the number of persons who actually deny the disability in this way and therefore refuse to participate in rehabilitation is small. (These cases have always involved strongly religious people who have viewed participation in rehabilitation as being in conflict with their belief in God's healing power.) However, most of the patients respond to the prognosis of paralysis with an assertion of will and strength; they will prove the physician wrong and walk out of the hospital. It is unfortunate that many professionals view such statements as evidence of a maladaptive coping process (denial) rather than as evidence of the powerful resources on which a person draws in times of adversity.

Caywood (1974) believes that information about the disability should be given in such a manner that it does not destroy the patient's hope that things might get better. He believes that hope need not interfere with the rehabilitation process but can provide a motive force to keep working despite the many frustrations of a treatment program. He affirms that hope is not the same as denial and should not be placed in the same category as a defense mechanism.

## THE STAGE THEORY OF ADJUSTMENT

The stage theory of adjustment has been discussed by many recognized experts in the field (Berger & Garrett, 1952; Gunther, 1969; Kerr & Thompson, 1974; Mueller, 1962; Siller, 1969), but little empirical evidence has been offered to substantiate the assertions of the theory. The stage theory suggests that the first reaction to the onset of disability is denial because of overpowering feelings of anxiety. This denial must be replaced by the second stage, depression, in order for adjustment to occur. The depression, according to the theory, represents a realistic and active mourning for the loss of valued functions and activities and has been considered to be the most important stage. The depression will be replaced by feelings of dependency and hostility, which must be worked through in order to reach the final stage, adjustment. It should be noted, however, that no evidence of stages has been presented in any of these discussions and that they represent the clinical impressions of the authors.

To date there is only one research project which has been designed to test the stage theory of adjustment. Dunn (1969) found that there was more variability than similarity in reaction to disability in his sample of 25 persons with spinal cord injury, and thus, there was no evidence of stages. Dexter (1981) has found no evidence of stages of adjustment in a longitudinal study of adjustment to spinal injury.

Since depression is considered to be a key element in the stage theory, research on this variable can shed some light on the validity of the theoretical formulations, which have dominated the field for 50 years. Taylor (1967) found that his sample of young men, who were within 3 months of onset of injury, displayed profiles on the Minnesota Multiphasic Personality Inventory (MMPI) that were similar to uninjured males on any university campus. There was little evidence of depression. Bourestom and Howard (1965) found that persons with spinal injury had the most benign MMPI profiles when compared to persons with rheumatoid arthritis and multiple sclerosis. Only mild depression was noted. Dinardo (1971) reports that persons with spinal injury who displayed evidence of depression in his study were independently rated as having the poorest adjustment to the disability in comparison to those who tended to suppress or repress their feelings. Lawson (1978) studied the incidence, severity, and pattern of depression in 10 persons with quadriplegia during their entire inpatient rehabilitation hospitalization. Four measures (self-report, biochemical, ratings by others, and behavioral) of depression were obtained on each person each day from admission to discharge. The results showed that depression was not a major factor in the overall pattern of behavior in his sample. There was evidence of mild depression 3 weeks after admission and just before discharge from the hospital. Otherwise, this sample did not display the pattern of depression which the stage theory predicts. Interestingly, there was variability among the measures of depression. The self-report and the biochemical and behavioral measures of depression were in agreement and showed few signs of depression except as noted above. In contrast, the ratings by staff consistently overestimated the amount of emotionality actually present. Thus, it is not surprising that the patients in this study reported that the most depressing influence was the staff expectation that they should be depressed.

Howell et al. (1981) evaluated 22 patients with spinal injury using a standardized interview and diagnostic process to assess the incidence of depression. All patients were within 6 months of injury. Depression was defined as a sustained and pervasive dysphoric mood accompanied by biological, behavioral, and cognitive symptoms. All interviews were conducted by psychiatrists. Their data showed that no patient suffered a major depressive episode following injury. Five patients showed mild depressive reactions defined as depressed affect, which was neither pervasive nor prolonged. Thus, they concluded that depressive disorders affect a minority of patients with traumatic spinal cord injury within the first few months of the accident. Depression does not appear to be a universal phenomenon, and when it does occur, it does not appear to be as severe or prolonged as expected.

What happens after discharge, when the full impact of the disability becomes apparent, has been studied by Richards (1986). He followed a group of spinal-injured persons into the community and found only moderate dysphoria and anger immediately after discharge and several months later; 1 year after discharge from the rehabilitation hospital these emotions had returned to the normal level.

These data should remind us that professionals as well as lay persons are vulnerable to perceiving something as evidence when it is consistent with expectations. The stage theory has been taught in professional curricula and repeated in most professional books in the section on adjustment to disability. Professional staff expect to see distress in persons with disability and consistently overreport its incidence and severity (Trieschmann, 1988). Taylor (1967) found that the staff of a highly regarded spinal injury center were no better than college fraternity men at predicting the incidence and severity of emotional reactions to spinal injury; both groups predicted such severe psychological reactions to the disability that a diagnosis of psychosis would have been necessary. In actuality, the sample of spinal-injured men had MMPI profiles similar to the average man on campus. Rosenthal (1966) has documented the existence of expectancy bias and has studied its effect on the results of experimental research. Thus, professionals need to examine some of their assumptions regarding the emotional reaction to disability. The notion that people must get depressed in order for adjustment to occur has not been substantiated. Yet numerous psychiatrists, psychologists, and social workers have encouraged disabled persons to express depressive statements and have labeled any tendency to resist this as denial. That, unfortunately, becomes a diabolical "Catch-22." If the disabled person admits to depression, this is a psychological problem that needs treatment; if the person claims not to be depressed, this is denial, a psychological problem which needs treatment.

The requirement of mourning has been described by Wright (1960) as the hypothesis that "when a person has a need to safeguard his values, he will either (1) insist that the person he considers unfortunate is suffering (even when he seems not to be suffering) or (2) devalue the unfortunate person because he ought to suffer and he does not" (Wright, 1960, 242–243). Wright also describes the requirement of mourning as the need to perceive the succumbing aspects rather than the coping aspects of living with a disability. Furthermore, there may be a discrepancy of expectation between the way the disabled person behaves and the way we expect him to behave. "He (the observer) may, for example, *alter the apparent reality* by doubting the evidence concerning the adequate adjustment of the person with a disability. Thus, he may feel that the person is shamming, simply acting *as though* he were managing, when actually he is not. He (the observer) may suppress evidence regarding the coping aspect of difficulties and highlight evidence bearing upon the succumbing aspects" (Wright, 1960, 73–74).

Is it possible that some of the publications that professionals have written reflect the "requirement of mourning"? Have professionals seen more distress and psychological difficulty than actually is present? Have they uncritically applied terms and theoretical concepts from the field of mental illness to describe the reaction to spinal injury? The onset of spinal cord injury is not a minimal event in one's life, yet many persons state that is not the most important thing that has happened to them during their lifetime (Corbet, 1980). Admittedly, it must be a most unpleasant experience, one that a person would prefer to avoid given the choice. But it seems clear that professionals have stressed the negative emotional aspects unnecessarily and underestimated the strengths and coping ability of people in crisis. Goldiamond (1976, 119–120) states:

When the professional refers to patients as "being unaware of," "being unrealistic about," or "repressing" their problems, it is the *professional* who is often being unrealistic. If I am not discussing pains, problems, and infections to which I am susceptible, it is not because I am unaware, unrealistic, or repressing. At times, I am painfully aware of them, and I mean that literally; I am sure other persons also do not discuss problems when they could. If they do not, in discussions with professionals, "face up" to these issues, it is because of the same good sense; they are facing, or trying to face, in a different direction, namely one that can help them program attainment of their goals.

## PERSONALITY AND SPINAL CORD INJURY

One of the fallacies of the stage theory of adjustment to disability was the implicit notion that there would be a sequence of emotional reactions which all persons would go through regardless of preinjury personality style. Research described in the previous section provides evidence to discredit a theory that assumes a homogeneous response to a disability. Professionals concur that there is no specific personality style or unitary reaction associated with any particular disability (Wright, 1960; Siller, 1969; Shontz, 1971; Cook, 1976; Trieschmann, 1988), yet there is a continuing tendency on the part of rehabilitation staff to make some global assumptions about the personalities of persons with spinal injury. Unfortunately, this is reinforced by research that attempts to describe the average response of persons with spinal injury to a personality test, such as the MMPI, for example.

There is no one personality style which is associated with spinal injury. "Persons with traumatically induced spinal injury comprise a heterogeneous population when they arrive for treatment: age, sex, cultural structure, education, marital status, experience in working and living, are as divergent as human nature itself. They will continue to be a heterogeneous group when they leave the centers, with one obvious difference: they will demonstrate a severe and probably permanent physical impairment" (Wilcox & Stauffer, 1972, 115). However, because a significant proportion of those with spinal injury are males, ages 15–29, there are certain behaviors often observed such as abuse of alcohol or drugs, physically active lifestyle, a tendency to engage in high-risk behavior, a tendency to prefer spontaneity rather than long-range planning, and a tendency to question authority. Such behaviors have a high base rate in the population of males in this age group, and thus, it is not surprising that a subset of this population acquires a spinal injury. But to ascribe these behaviors to persons with spinal injury rather than to the population from which they come is fallacious. This group is also only one part of the population of persons with spinal injury, and there is no simple way to characterize the remainder.

Several previously mentioned studies have looked at the average MMPI profile of persons with spinal injury. Bourestom and Howard (1965) found that a group with spinal injury was only mildly depressed in comparison to a group of persons with rheumatoid arthritis and multiple sclerosis. The shapes of the profiles of these three groups were somewhat similar, but the elevation of the profiles of the latter two groups was higher. Taylor (1967) found that the average MMPI profile for his group of newly injured young men was similar to that of men on college campuses: slight elevations on scale 4 (Pd) and scale 9 (Ma), indicating independence, assertiveness, and much energy. Thus, this is an excellent example of the influence of demographic variables on base rates of behavior for this group.

Nevertheless, the question remains: If there is no unitary response to spinal injury (no spinal-injury personality nor uniform stages of adjustment), what do we learn from descriptions of the average response on any personality test or assessment device? Why should we anticipate a homogeneous reaction to spinal injury when the only feature these persons have in common is the physical disability? This homogeneity of response is an implicit assumption whenever one considers average MMPI profiles or average scores on other measures. Trieschmann and Sand (1971), studying terminal renal patients, observed their intellectual and personality response to the process of renal failure. The average MMPI profile was similar in shape to those in the Bourestom and Howard study (1965) and showed a very low correlation to measures of kidney function. However, when the profiles were sorted according to type of response to the crisis, five different reaction types emerged. It was hypothesized that a person's response to a major life crisis would be similar to one's typical response to severe stress and that people differ in these response styles depending on their precrisis personality. Thus, five different profile types were obtained, none of which was similar to the total-group average profile. There were vast individual differences that were obscured by the averaging process; the total average profile was not descriptive of anyone in the study sample. As a result, averaging MMPI profiles or other tests of personality for heterogeneous groups of people must be challenged. It would be much more fruitful to study the incidence of a particular personality trait and its association with specific behaviors or to categorize persons according to certain behaviors and study the incidence of certain personality traits.

For example, Fordyce (1964) studied a sample of males with spinal injury who had been categorized into two groups: those whose disability occurred as a result of their own imprudent behavior and those whose injury occurred through accident with no evidence of imprudence on their parts. He found that the group categorized as imprudent scored higher on scales 3 (Hy) and 4 (Pd) of the MMPI than the prudent group. In a similar study, Kunce and Worley (1966) tested two groups on the Strong Vocational Interest Blank (SVIB). One group was composed of persons who were active agents in their accidents and the other group consisted of persons who were passive agents. Those who were active agents in their accidents scored higher on the aviator key of the SVIB, which is often interpreted as showing evidence of adventurousness, boldness, and assertiveness. However, the issue of base rates must be considered in both of these studies since the results may reflect the influence of age and sex variables. Nevertheless, they represent an approach to the study of persons with spinal injury that is based on the premise of individual differences.

## MOTIVATION

Motivation is an important factor in the adjustment to spinal cord injury, and it has received considerable attention in the literature. However, it is a summary term, which we use to describe all of the features that determine whether the person will incorporate the teachings of rehabilitation into his or her lifestyle. Traditionally, we have assumed that motivation is exclusively a characteristic of the person; recently, however, there has been an increasing amount of evidence to suggest that the definition of motivation as an internal drive state is too limited a view of the situation (Fordyce, 1971). Rather, we have evidence that the use of operant techniques on "unmotivated" persons changes their behavior so that it is similar to that of "motivated" persons. Thus, we have begun to look at the environment as a critical element in the assessment of the person's motivation (Trieschmann, 1988). In this approach, unmotivated persons are the ones for whom we have no reward for which they will work. The focus is external to the person and is on rewards and punishments in the environment.

Seligman (1975), however, has proposed that merely having a powerful reward available in the environment may not be sufficient if the person has learned that there is no contingency between his behavior and the outcomes. He calls this learned helplessness. Spinal injury may be a dramatic example of a state in which learned helplessness can occur (Trieschmann, 1988). With the sudden onset of paralysis and the loss of control over one's life that hospitalization entails, certain people may be vulnerable to learning to be helpless, to believing that there is nothing that they can do to improve their situation.

There are individual differences in susceptibility to learned helplessness, and the locus-of-control dimension may account for a significant part of the variance. Locus of control is an expectancy that one brings into a learning situation; it may be defined in terms of the expectation that one can control the rewards the world has to offer. Those with an internal locus of control believe that their behavior will be rewarded if they work hard, whereas those with an external locus of control believe that fate, luck, and powerful others control the rewards of the world. Seeman and Evans (1962) reported that those with an internal locus of control were more interested in gaining knowledge of their disease than those with an external locus of control, because the "internals" saw such knowledge as useful in controlling their lives. Dinardo (1971) found that internals had a better self-concept and were less depressed after spinal injury than "externals." Swenson (1976) provides the most complete research to date on the relevance of locus of control to adjustment to spinal injury. Internals were found to have spent less time in the hospital as a result of preventable medical problems; spent more time in work activities in the home, in educational activities, and in time outside the home; and spent more time in a combination of education, paid employment, and community work. There was no correlation between locus of control and level of disability. Thus, although a person with quadriplegia may have less physical control over his life, this does not change the generalized expectancy of control over the rewards and satisfactions of life.

Anderson and Andberg (1979) found that those with the highest incidence of time lost because of decubitus ulcers showed the least acceptance of responsibility for care of their skin. Even with high levels of disability, acceptance of responsibility in a conceptual sense was the paramount factor in prevention of pressure sores. Kemp and Vash (1971) found that having a large number of goals was highly correlated with productivity and successful adjustment to spinal injury. Although they did not measure locus of control, it is interesting to wonder if internals would report more goals than externals or if the type of goal would differ. The more productive persons in their study reported goals in the vocational and familial-interpersonal categories, whereas less productive persons had goals that were avocational and physical in nature.

Although management of psychosocial issues will be discussed later, it is important to note here that the management of motivational deficits will vary depending upon whether one applies the medical model of rehabilitation or the learning model. Within the medical model, lack of motivation is viewed as the patient's problem, and therefore the psychologist is asked to fix this problem, preferably in the psychologist's office, using counseling or psychotherapy. Unfortunately, these treatments have been unsuccessful at improving motivational level (Diamond et al., 1968). But in the learning model of rehabilitation, motivational deficits are viewed as an environmental problem, and the solution is found in modifying the patient-environment interaction (Trieschmann, 1988; Trieschmann & Willems, 1980). This usually means that the rehabilitation staff need to look at their own behavior and the operational policy of the team to seek ways to produce the desired behavior from the disabled person. In the medical model, the patient is the failure; in the learning model, the program is the failure. Note that the learning model does not imply that the patient has no responsibility to bring a willingness to cooperate to a therapy program. It does imply that the issue of motivation is complex and usually involves an interaction of the person and the environment in which he or she is asked to perform.

## MANAGEMENT STRATEGIES

Use of the learning model of rehabilitation requires a recognition that the behavior of the person with spinal injury will reflect the interaction of the person's personality with environmental influences. As a general principle of management, if the patient's behavior is being deemed inappropriate, first analyze whether the desired behavior is truly "appropriate." Rehabilitation staff and hospital personnel have a great affection for patient compliance with requests, routines, policies, and orders. Compliance, however, may not always be in the best interests of the disabled person in the long run, because compliance with externally imposed routines does not teach independence, problem solving, and coping with a disability. Second, if the person is not behaving in a desired way, examine the impact of the person's environment, and try to assess whether the behavior is in opposition to or consonant with that environment.

An example of behavior in opposition to the environment is the noncompliance with nursing ward routine by young, spinal-injured men. As a group, they are independent, aggressive, active, and not overly respectful of authority. Therefore, a rehabilitation routine that emphasizes strict rules and regulations, strictly enforced time schedules, and many prohibitions is the ideal environment to provoke opposition rather than compliance. Naturally, every ward needs certain procedures and regulations to promote the well-being of all patients, but the attitude with which the staff convey these policies can influence the compliance rate. Furthermore, a rehabilitation staff may inadvertently develop such a strict and rigid program for certain "problem" cases, that the only way in which the disabled person can exert his or her independence is by not doing what is expected. Such situations lead to an escalation of the conflict with noncompliance, leading to further authoritarianism, which in turn leads to further noncompliance.

The resolution of such situations requires a mediator, often the psychologist, who asks the patient and staff to sit down together and negotiate a new mode of interaction. This usually involves both parts giving in and affirming an agreement to work together. As long as the team and the patient agree on the ultimate goal—the maximum function of the patient and discharge from inpatient status—they can plan steps to accomplish this goal. The confrontation needs to be ended and the personal irritations defused in order to promote cooperation.

An example of behavior in consonance with the environment is the nonparticipation in activities of daily living by a man who belongs to a certain culture in which women are expected to take care of "incapacitated" individuals. The rehabilitation center may make it perfectly clear that the man is expected to learn to dress, bathe, groom, and feed himself. Nevertheless, the man may not see the necessity of learning these activities because they are women's work. An associated concept within such cultures is the expectation that the disabled man's productive life is at an end and that the family will take care of him for the rest of his life. While we may not agree that productivity needs to end with the onset of disability, it is unrealistic to invest great effort in a program that places the person in total conflict with his or her family and culturally defined role. This is more likely to occur in recent immigrants to the United States or Canada who have not assimilated Western norms and culture. It is also likely in much older citizens; older men frequently do not see the need to expend the energy required to function at their maximal level and will settle for less independence in exchange for assistance. In such cases, the team needs to make a realistic assessment of the person's culture and the behaviors that have a reasonable probability of being continued in the home environment. In order to plan an appropriate program, the goals and wishes of the disabled person and the family must be understood.

Young men with spinal injury are active, as a group, and their major way of relating to the world has been markedly changed by the spinal injury. Even if they talk about "walking out of here," this is not necessarily denial and should not be challenged verbally. To "not want to live like this" is not necessarily suicidal ideation but merely a statement of fact, particularly since the newly injured person has no information or understanding about what life will be like with a physical disability. This understanding will take several years. Meanwhile, a rehabilitation program will be planned, which will help the person find meaningful and productive activities to offset the hard work and frustration that disability entails.

Typically, people do not get deeply depressed soon after the injury. Depression should be defined as insomnia, lack of appetite, and psychomotor retardation of at least 3 days duration. In the rare case in which this state occurs, antidepressant medication on a temporary basis may help to initiate some of the behaviors needed to fully participate in a rehabilitation program. Depression at this stage is not a good prognostic sign; therefore, a counseling and rehabilitation program that emphasizes activity and productivity is certainly indicated. Most persons with spinal injury will be very unhappy about their state, and loving empathy is the treatment of choice. At the same time, they should not be encouraged to dwell on losses and sad feelings but rather to look at the assets they still have and to do the best they can in the program. However, the physical rehabilitation program should be considered as only the first stage in a multifaceted endeavor leading to resumption of a satisfactory life in one's community.

Because motor activity has been such a significant part of these persons' lives, the rehabilitation program should seek ways to channel this energy in appropriate directions. In-hospital sports programs and extra physical therapy can help burn up the emotional tension that sudden onset of disability produces. If these young men have suitable opportunities to expend this energy physically (which is their preferred mode), they are less likely to look for ways to disrupt hospital routine because of boredom. Furthermore, physical fatigue produces a more satisfying sleep than that induced by medication.

Typically, rehabilitation programs place most emphasis on teaching activities of daily living, bladder and bowel management, and mobility techniques. But these are the skills which the average 5-year-old child has acquired. Successful adult life requires a variety of sophisticated interpersonal skills, which the onset of a physical disability makes even more necessary. However, these skills, plus community-living skills, receive much less emphasis or are essentially ignored in most rehabilitation centers. Thus, a multistage rehabilitation program is essential, with acute-injury management, physical restoration, and community-living training having equal importance in the long-term program. With such an approach we can assist persons with spinal injury to live meaningful and productive lives and to be psychosocially integrated into their own communities.

## SUMMARY

The population of persons who acquire a spinal injury is predominantly young, male, and active. These people are in the midst of mapping out courses that would shape their adult lives, and thus the physical disability serves as a major interruption and often as a major refocusing of life activities. There is, however, no evidence of a spinal-injury personality, and each person will cope with the disability in a manner consistent with his or her preinjury style of coping with stress.

Although earlier literature emphasized the stage theory of adjustment, current research suggests that people do not go through a reliable sequence of stages but rather follow their unique style of coping. Depression, the evidence demonstrates, does not occur with great frequency and, contrary to the stage theory, is not a necessary precursor to adjustment. On the contrary, there is evidence that early depression following a disabling injury is not a good prognostic sign. Depression, however, should be differentiated from unhappiness just as denial should be differentiated from hope.

In order to plan a proper rehabilitation program that will promote independence and successful community living, the models of rehabilitation should be examined to use the proper strategy for the proper task. The medical model is appropriate during the acute-injury phase but not during rehabilitation. Rather, at that point the learning model should be used, because it emphasizes the powerful and pervasive influence of the environment on the outcome of our rehabilitation efforts.

## REFERENCES

Anderson, T. & Andberg, M. (1979). Psychosocial factors associated with pressure sores. *Archives of Physical Medicine and Rehabilitation, 60,* 341–346.

Berger, S. & Garrett, J. (1952). Psychological problems of the paraplegic patient. *Journal of Rehabilitation, 18,* 15–17.

Bourestom, N. & Howard, M. (1965). Personality characteristics of three disability groups. *Archives of Physical Medicine and Rehabilitation, 46,* 626–632.

Braakman, R., Orbaan, J. & Dishoech, M. (1976). Information in the early stages after spinal cord injury. *Paraplegia, 14,* 95–100.

Caywood, T. (1974). A quadriplegic young man looks at treatment. *Journal of Rehabilitation, 49,* 22–25.

Cook, D. (1976). Psychological aspects of spinal cord injury. *Rehabilitation Counseling Bulletin, 19,* 535–543.

Corbet, B. (1980). *Options: Spinal cord injury and the future.* Newton Upper Falls, Mass.: National Spinal Cord Injury Foundation.

Dexter, W. (1981). Personality factors associated with adjustment to disability. Paper presented at the Annual Convention of the American Psychological Association, August, at Los Angeles, Calif.

Diamond, M., Weiss, A. & Grynbaum, B. (1968). The unmotivated patient. *Archives of Physical Medicine and Rehabilitation, 49,* 281–284.

Dinardo, Q. (1971). Psychological adjustment to spinal cord injury. Ph.D. diss., University of Houston, Houston, Tex.

Dunn, D. (1969). Adjustment to spinal cord injury in the rehabilitation hospital setting. Ph.D. diss., University of Maryland.

Fordyce, W. (1964). Personality characteristics of men with spinal cord injury as related to manner of onset of disability. *Archives of Physical Medicine and Rehabilitation, 45,* 321–325.

Fordyce, W. (1971). Behavioral methods in rehabilitation. In W. Neff (Ed.), *Rehabilitation psychology.* Washington, D.C.: American Psychological Association.

Goldiamond, I. (1976). Coping and adaptive behaviors of the disabled. In G. Albrecht (Ed.), *The sociology of physical disability and rehabilitation.* Pittsburgh: University of Pittsburgh Press.

Gunther, M. (1969). Emotional aspects. In R. Reuge (Ed.), *Spinal cord injuries.* Springfield, Ill.: C. C. Thomas.

Harris, P., Patel, S., Greer, W. & Naughton, J. (1973). Psychological and social reactions to acute spinal paralysis. *Paraplegia, 11,* 132–136.

Hohmann, G. (1975). Psychological aspects of treatment and rehabilitation of the spinal injured person. *Clinical Orthopedics, 112,* 81–88.

Howell, T., Fullerton, D., Harvey, R. & Klein, M. (1981). Depression in spinal cord injured patients. *Paraplegia, 19,* 284–288.

Kemp, B. & Vash, C. (1971). Productivity after injury in a sample of spinal cord injured persons: A pilot study. *Journal of Chronic Disease, 24,* 259–275.

Kerr, W. & Thompson, M. (1974). Acceptance of disability of sudden onset in paraplegia. *Paraplegia, 10,* 94–102.

Kunce, J. & Worley, B. (1966). Interest patterns, accidents, and disability. *Journal of Clinical Psychology, 22.* 105–107

Lawson, N. (1978). Significant events in the rehabilitation process: The spinal cord patients' point of view. *Archives of Physical Medicine and Rehabilitation, 59,* 573–579.

Mueller, A. (1962). Psychologic factors in rehabilitation of paraplegic patients. *Archives of Physical Medicine and Rehabilitation, 43,* 151–159.

Richards, J. S. (1986). Psychological adjustment to spinal cord injury during first year after discharge from the rehabilitation hospital. *Archives of Physical Medicine and Rehabilitation, 67,* 362–365.

Rosenthal, R. (1966). *Experimenter effects in behavioral research.* New York: Appleton-Century-Crofts.

Seeman, M. & Evans, J. (1962). Alienation and learning in a hospital setting. *American Sociological Review, 27,* 774–782.

Seligman, M. (1975). *Helplessness: On depression, development, and death.* San Francisco: W.H. Freeman.

Shontz, F. (1971). Physical disability and personality. In W. Neff (Ed.), *Rehabilitation psychology.* Washington, D.C.: American Psychological Association.

Siller, J. (1969). Psychological situation of the disabled with spinal cord injuries. *Rehabilitation Literature, 30,* 290–296.

Swenson, E. (1976). The relationship between locus of control expectancy and successful rehabilitation of the spinal cord injured. Ph.D. diss., Arizona State University, Tempe, Ariz.

Taylor, G. (1967). Predicted versus actual response to spinal cord injury: A psychological study. Ph.D. diss., University of Minnesota, Minneapolis, Minn.

Trieschmann, R. (1987). *Aging with a disability.* New York: Demos Publications.

Trieschmann, R. (1988). *Spinal cord injuries: Psychological, social, and vocational rehabilitation.* New York: Demos Publications.

Trieschmann, R. & Sand, P. (1971). WAIS and MMPI correlates of increasing renal failure in adult medical patients. *Psychological Reports, 29,* 1251–1262.

Trieschmann, R. & Willems, E. (1980). Dehavioral programs for the physically disabled. In D. Glenwick & L. Jason (Eds.), *Behavioral community psychology: Progress and prospects.* New York: Praeger Publishing Corp.

Wilcox, E. & Stauffer, S. (1972). Follow-up of 423 consecutive patients admitted to the Spinal Cord Center, Rancho Los Amigos Hospital, 1 January to 31 December, 1967. *Paraplegia, 10,* 115–122.

Wright, B. (1960). *Physical disability: A psychological approach.* New York: Harper & Row.

Young, J., Burns, P., Bowen, A. & McCutchen, R. (1982). *Spinal cord injury statistics: Experience of the regional spinal cord injury systems.* Phoenix: Good Samaritan Hospital.

Zubek, J. (Ed.). (1969). *Sensory deprivation: Fifteen years of research.* New York: Appleton-Century-Crofts.

# 8
# Psychosocial Implications and Treatment of Persons with Multiple Sclerosis

*Sheldon Berrol*

Multiple sclerosis, the predominant model for progressive adult neurologic disorders, has an average age of onset of between 20 and 40 years. Thus, the impairment of an individual's functional abilities arises at a point when his or her life situation is developing stability and the future appears most promising. The disabling consequences of the disease process are associated with characteristic emotional responses. Unfortunately, our understanding of these responses has been obscured over the past 100 years by a lack of understanding of the disease process and its sequelae.

## HISTORICAL PERSPECTIVE

Charcot (1877) first suggested that intellectual and emotional faculties are totally blunted in multiple sclerosis (MS), yet further identified euphoria and morbid optimism as characteristics of the "multiple sclerosis personality." The following century of clinical research was devoted to validating his impressions. Sugar and Nadell (1943) attempted to ascertain if euphoria was a distinct concomitant of MS and concluded that patients demonstrated a sense of well-being that was not consistent with their physical condition. Except for this single study, however, no other researchers since 1943 have been able to duplicate Charcot's conclusions.

Research during this later period seemed instead to identify a characteristic waxing and waning of symptomatology. This led to the perception that the pathologic process went through phases of amelioration, during which there was, in addition to symptomatic improvement, a distinct improvement in the underlying pathologic substrate alternating with phases of deterioration. There is now, however, considerable question as to whether exacerbation and remission are in fact hallmarks of MS. Current research suggests that "it seems logical now to consider the clinical manifestations of MS as the externally observable part of an underlying continuous process" (Poser, 1980).

## SYMPTOM ONSET

Initially, symptomatology is vague and may lack sufficient evaluative corroboration so that several years generally pass before a definitive diagnosis can be established. Weakness as a symptom is difficult to authenticate. Paresthesia is also a complaint that relies solely upon the perception of the patient. Pain or numbness, without the ability to relate them to a specific localizing pathology, and nonspecific visual problems in the face of reasonably good visual acuity, are vague symptoms that perpetuate the diagnostic difficulty. During this 2–3-year period of delay, patients are frequently told that they are suffering from the stress and anxiety of their job, a relationship, their abilities, and the like, resulting in anxiety, depression, and a wide variety of other psychiatric disturbances. The end product of such communications is considerable self-doubt, marked reduction in ego strength and the production of behavioral responses that may become difficult to deal with as the disease progresses. Indecisiveness among treating physicians is without doubt a major factor contributing to the level of stress and altered coping ability of the patient. The patient, having been given a variety of diagnostic alternatives, frequently selects from among the less disabling choices. Professionals, not uncommonly, see in this an example of denial rather than a reflection of their inability to establish the diagnostic entity early.

## MULTIPLE SCLEROSIS PERSONALITY

Years of research have been expended in an effort to categorize the "multiple sclerosis personality" (Cohen, 1962; Harrower & Kraus, 1951). It appears now that there is no foundation for the concept that specific disease states produce specific emotional disorders. Rather, it seems quite apparent that any progressive, chronic disabling condition will produce a variety of behavioral and emotional responses based on the person's coping strategies, which have developed under the influence of environment, peer relationships, ego strength, and other variables (LaRocca et al., 1983). These coping strategies are not at all maladaptive. What is misdiagnosed as euphoria, for example, may be a false sense of optimism allowed by the hesitancy of health professionals to establish an early diagnosis. It is merely part of a normal coping strategy. Though sometimes mislabeled as denial, it must be remembered that the health care professionals originally offered a choice of diagnoses and the patient has selected the most hopeful one. The emotional lability so often mentioned in the literature is rarely seen in clinical practice except in the most advanced cases.

Hysteria has repeatedly been suggested as a component of the "multiple sclerosis personality," most frequently when the symptomatology seemed inappropriate to the physical findings. These investigators have failed to consider that although the perception of functional loss may accentuate an emotional response, the loss really did occur (Peyser et al., 1980a, 1980b). Miller et al. (1965) felt that one-third of patients with MS who complained of sexual dysfunction had overriding emotional problems. At the same time, they did not doubt the organic basis of the dysfunction. What may appear as an overreaction may be a reflection of that individual's inherent concept of organ value, self-worth, and gender identity. Other studies (Minderhoud et al., 1984) have concluded that not only are sexual disturbances a very frequent symptom of the disease, but that there is clearly an organic basis. In order to properly counsel their patients, physicians must have a comprehensive understanding of the disease process and be able to filter out any observer bias toward functional etiology.

## COGNITIVE DEFICITS

Canter (1951) and Shontz (1955), using profiles on the Minnesota Multiphasic Personality Inventory, found clinically significant elevations in the so-called neurotic triad: hypochondriasis, depression, and hysteria. They suggested that depression, anxiety, and bodily concern were significant in patients with MS. In fact, regardless of diagnosis, loss of bodily function frequently results in depression, anxiety, and bodily concern. It is naive to expect that all patients, regardless of their unique personalities and symptom complexes, will react in the same fashion and fit into a distinct, single pattern. Psychological adjustment depends upon the interplay of the central nervous system, the patient's premorbid personality pattern, and a wide variety of external variables.

A number of investigators have attempted to identify cognitive and intellectual deficits in the MS patient (Jambor, 1969; Lincoln, 1981; Peyser et al., 1980a). Surridge (1969) found that almost two-thirds of the patients he evaluated demonstrated intellectual deterioration, chiefly deficits in memory and conceptual thinking, and one-fourth demonstrated depression. He concluded that euphoria was strongly associated with intellectual deterioration and denial, and further, that intellectual deterioration and personality change were symptomatic of the disease—directly because of damage to the central nervous system. He recommended that adequate assessment of intellectual function become standard in the management of MS. Peyser et al. (1980a) stated that cognitive impairment is often not detected on routine examination, but postulated that it may occur early in the course of the disease. Some 54.7% of the subjects demonstrated impaired cognitive ability that was not related to depression or fatigue but was directly caused by otherwise quiescent plaques in the subcortical white matter. Since the subjects previously had thorough neurologic evaluations, it was evident that more comprehensive procedures were necessary to detect these abnormalities.

These conclusions were presaged by Young et al. (1976) who described five MS patients, with mental change as an early and prominent symptom. In marked contrast to earlier thinking that mental changes were rare early in the course of MS, they postulated that such early changes resulted from widespread plaques of demyelination confined not just to the brain stem but present throughout the cerebral hemispheres. Peyser et al. (1980a), after convincingly reviewing neuropathologic and neuropsychologic data and reflecting on the frequency of cognitive deficits in MS patients, urged the inclusion of cognitive evaluation in the capital disability rating scale. They further proposed that a specific screening battery for cognitive disturbance be developed for administration to individuals with MS.

It currently appears that different degrees of cognitive impairment may be an almost invariable concomitant of MS and that in a substantial number of cases, it may occur rather early (Young et al., 1976). It may well be that what appears as psychological denial may reflect a level of impaired awareness, much as one sees in a population of brain-injured persons. One can then reasonably argue that such denial is different from that which occurs when full perceptual abilities are intact and either a conscious or unconscious rejection of information occurs.

## EMOTIONAL DISTURBANCES

Dalos et al. (1983) evaluated the prevalence and nature of emotional disturbances in MS patients and compared them with spinal cord–injured controls. They found a 90% prevalence of emotional disturbances in patients whose conditions were exacerbating or progressing and 39% in patients with stable conditions, compared with 12% in the spinal cord–injured control group. They felt that the MS patients in remission demonstrated a higher level of somatic complaints, anxiety, and social dysfunction than symptoms of depression. If we evaluate these findings in the light of a disease state that produces constant loss of function, then these patients may have a firm foundation for their complaints. While MS patients may go into remission, it is more likely they will be more disabled than they were during the previous remission. Thus, one cannot assume that an MS patient in remission carries the same degree of emotional stability as the spinal cord–injured patient.

Maybury and Brewin (1984) attempted to evaluate the psychological factors that determined a patient's adjustment to MS. They found that in spite of considerable individual variation in knowledge about MS, no correlation between knowledge and psychological adjustment existed. They postulated a variety of reasons, suggesting that some people who are distressed about their condition may deliberately seek more information, while others may deliberately avoid it. (They failed to suggest, however, that the results may in part reflect our inability to provide meaningful reassurance to the patient.) They further felt that patients were unlikely to know what help to expect from their physicians or what to expect during a remission. They concluded that improved adjustment may not follow from a "textbook" knowledge of the facts but may require knowledge that can be put to some practical use. They suggested that patient education should focus on practical matters of direct relevance to patients trying to manage their condition.

## STRESS

A wealth of mythology has been created regarding stress and MS. Specific perspectives in the past have been that stress causes MS, that stress triggers

MS, and that stress makes MS worse (Weinstein, 1970). Many early investigators felt that the pathogenesis of the MS plaque required stress as an initiating factor for it to develop. A wide variety of stressful conditions were considered causative, including a death in the family, a new job, moving to a new home, or environment (Mei-Tal, 1970). These conclusions were drawn from a variety of poorly designed studies.

Grinker et al. (1950) studied 26 cases. In 6 of these cases they were able to identify a major life-situation change immediately preceding the development of symptoms that led to the diagnosis of MS. They concluded that these individuals must have had an immature personality, one that could not cope with a significant life change, and so in response, they developed MS. This conclusion is open to considerable question, I believe, since 20 of the cases did not fit the schema the authors themselves postulated. Baldwin (1952) compared 34 cases of MS with 34 control cases of various neurologic etiologies. He concluded that both groups suffered equally in terms of emotional trauma. He stated further, however, that MS patients seemed to have less stability in their lives. Although not stated in the study, it would be interesting to know whether the neurologic conditions in the controls were progressive. The state of stability versus progression in a disease process may significantly affect one's ability to cope.

The "stress as causation" theory was also promoted by Langworthy (1948), who argued that if stress was a major factor in the development of the MS plaque, then psychotherapy should be used as prophylaxis for preventing MS. Were his recommendations heeded, there would be little time for psychiatrists to do anything else. Pratt (1951) examined 100 patients with MS and compared them to a control group of 100 patients with other neurologic conditions. Thirty-eight cases of MS and twenty-six controls had evidence of severe stress prior to the onset of their symptoms. In drawing his conclusions, he felt that eight of the MS cases demonstrated such a strong cause and effect relationship, that there would be no doubt in his mind about the role of stress in the pathogenesis of the disease process.

Attention of the lay press has been drawn once again to stress as an etiologic factor with the development of MS symptomatology by Richard Queen, one of the hostages at the American embassy in Iran a few years ago. Stress and MS must be viewed in the perspective of a disease process characterized by apparent exacerbations and remissions of symptomatology. The disease is, in fact, a lifelong crisis. Most illnesses produce an initial crisis without the continual need for physical and emotional adaptation required of the MS patient. It has been frequently claimed that stress plays a major role in the exacerbation of MS symptomatology. Perhaps we should consider the fluctuations of any disorder that produces sudden loss of or interference with life functions to be substantially stressful. Too often, the MS patient is told to avoid stress. But how, in modern society, does one accomplish this? The constant admonition to avoid stress burdens the patient with a level of responsibility and subsequent guilt that is unreasonable. In reality, the stress of MS is the stress of any serious progressive illness. Additionally, the stress of MS is in large part a consequence of the uncertainty the individual faces with respect to prognoses. The constant and progressive loss of function impairs self-esteem, producing incapacitation, helplessness, and a feeling of abandonment. Such thoughts adversely affect patients and their relationships with family and significant others. A not uncommon sequela is the assumption of guilt by the family, which further complicates intrafamilial dynamics and leads once again to stress. At the present time, the relationship between multiple sclerosis and stress remains, at best, ambiguous; the role of stress in precipitating symptomatology must be considered minor.

## DENIAL

Denial can be a useful survival mechanism. If the patient is coping well with the disability, then there may be no advantage to psychological intervention. It must be recalled that denial is reinforced by a lengthy delay in the establishment of a diagnosis, the vague symptomatology, the uncertainty of outcome, and the almost invariable psychiatric diagnosis that labels the patient at the onset. The development of periods of remission may, in the early years, reinforce the patient's denial of the ultimate diagnosis. A further complicating issue is the extreme limits of function that are occasionally and inaccurately placed upon the patient, such as, "You'll be in a wheelchair in 2 years," "You'll become impotent in 6 months," and the like. It appears that in general the functional ability of the patient remains better than the health professional anticipates. Baldwin (1952) noted that there is a tendency for denial to become substantially reduced as neurologic involvement and its resultant disability increase with time. One-third of MS patients will continue to live a totally normal life-style for up to 5 years after the establishment of a diagnosis, two-thirds will remain ambulatory for at least 25 years, and one-fourth will continue to work 30 years after being diagnosed (Bauer, 1978; Percy et al., 1971). In spite of the recent literature suggesting a persistent development of the pathogenic process, its rate of progress varies substantially.

## MEDICATION

The MS patient frequently imbibes a variety of medications for bladder management, control of spasticity, pain, constipation, and management of depression and anxiety. Each medication must be individually evaluated for its potential to exacerbate symptoms or adversely interfere with an already marginal function. Thus, anticholinergic medication such as tricyclic antidepressants or antihistamines may result in bladder retention in the presence of prostatic hypertrophy, further potentiating any emotional-management problems. Additionally, this group of medications has the potential to impair memory and learning. The resulting cognitive deficit could present itself as a marked functional deterioration. On the other hand, during an exacerbation when incontinence becomes a factor, then the risks of untoward reactions of anticholinergics may be warranted.

Prescriptions for upper motor neuron spasticity are few, and each one has potential undesired effects on mental function, personality, or strength. An individual on any of these medications may yield substantial abnormalities on intellectual or personality evaluations. If functional gains are evident, then the use of these medications is clearly indicated. However, if psychologic deterioration results, then an analysis of the medication regime is called for prior to the pronouncement of an emotional or psychiatric diagnosis.

## SUMMARY

Cognitive impairment may occur quite early in the course of MS, which may significantly affect the patient's emotional level. However, uncertainty regarding diagnosis, the paucity of information regarding loss of functional ability, and the increasing pressures to maintain a stable life-style in the presence of decreased functional ability may be of greater significance in the production of emotional symptoms. The patient's coping mechanisms must not be discounted, even though they may at times be less than effective. These intrinsic systems of management need to be enlarged upon and made more effective.

Dealing with the stress and emotional issues associated with MS requires an overall knowledge of the disease. The information must be provided not only to the patient but also to families and significant others. The knowledge base must be broad enough to incorporate each factor of physical and psychological function that is or may be affected by the disease process.

The health care team and the patient's significant others must develop increased sensitivity to the sense of loss and powerlessness that the patient suffers. The reality of the loss must be acknowledged before the building blocks of emotional competency can be reestablished. Individuals must be provided opportunities to verbalize their feelings of anger, frustration, and depression. Both individualized and group processing is essential, and the counselor must be well informed regarding not only MS but the issues surrounding chronic disability. Comprehensive evaluation of the individual's environment must be established so that adequate modifications and compensations can be made to allow maximal function. MS patients must be aided in setting realistic goals and expectations based on their strengths, limitations, and the reality that they face.

There is a sufficient body of knowledge available to allow us to rethink traditional perspectives on the emotional and psychological consequences of MS. While it now appears that the course of the disease is continual rather than sporadic, the potential for the patient's full integration into the home and the community is greater than previously assumed. Increasing the functional capabilities of each individual promises to substantially reduce the negative emotional aspects of the disease process. Psychologic intervention, within the context of a multidisciplinary team approach, should be based upon individual needs rather than presupposed generalizations.

## REFERENCES

Baldwin, M. V. (1952). A clinico-experimental investigation into the psychologic aspects of multiple sclerosis. *Journal of Nervous and Mental Disease, 115,* 299–342.

Bauer, H. J. (1978). Problems of symptomatic therapy in multiple sclerosis. *Neurology, 28* (9), 8–20.

Canter, A. H. (1951). Direct and indirect measures of psychological deficit in multiple sclerosis. *Journal of Genetic Psychology, 44,* 27–50.

Charcot, J. M. (1877). *Lectures on diseases of the nervous system.* London: New Sydenham Society.

Cohen, A. (1962). Personality aspects of multiple sclerosis. A review of the literature from 1950 through 1961. *Journal of Rehabilitation. 28*(3), 18–20, 50–51.

Dalos, N. P., Rabins P. V., Brooks, B. R. & O'Donnell, P. (1983). Disease activity and emotional state in multiple sclerosis. *Annals of Neurology, 13,* 573–577.

Grinker, R. R., Ham, G. C. & Robbins, F. P. (1950). Some psycho-dynamic factors in multiple sclerosis. *Research Publications of the Association of Nervous and Mental Disease, 28,* 456.

Harrower, M. R. & Kraus, J. (1951). Psychological studies on patients with multiple sclerosis. *Annals of Neurology, 66,* 44–51.

Jambor, K. L. (1969). Cognitive functioning in multiple sclerosis. *British Journal of Psychiatry, 115,* 765–775.

Langworthy, O. R. (1948). Relation of personality problems to onset and progress of multiple sclerosis. *Archives of Neurology and Psychiatry, 59,* 13.

LaRocca, N., Kalb, R. & Kaplan, S. R. (1983). Psychological changes. In L. C. Scheinberg (Ed.), *Multiple sclerosis, a guide for patients and their families.* New York: Raven, 175–194.

Lincoln, N. B. (1981). Discrepancies between capabilities and performance of activities of daily living in multiple sclerosis. *International Rehabilitation Medicine, 3,* 84–88.

Maybury, C. P. & Brewin, C. R. (1984). Social relationships, knowledge, and adjustment to multiple sclerosis. *Journal of Neurology, Neurosurgery, and Psychiatry, 47,* 372–376.

Mei-Tal, V., Meyerowitz, S. & Engel, G. L. (1970). Role of psychological process in a somatic disorder: Multiple sclerosis. *Psychosomatic Medicine, 32,* 67–86.

Miller, H., Simpson, C. A. & Yeats, W. K. (1965). Bladder dysfunction in multiple sclerosis. *British Medical Journal, 1,* 1265–1269.

Minderhoud, J. M., Leemhuis, G., Kremer, J. et al. (1984). Sexual disturbances arising from multiple sclerosis. *Acta Neurologica Scandinavica, 70,* 299–306.

Percy, A., Nobrega, F., Okazaki, H., Glattre, E. & Kurland, L. T. (1971). Multiple sclerosis in Rochester, Minnesota: A 60-year appraisal. *Archives of Neurology, 25,* 105–111.

Peyser, J. M., Edwards, K. R., Poser, C. M. & Filskov, S. B. (1980a). Cognitive function in patients with multiple sclerosis. *Archives of Neurology, 37,* 577–579.

Peyser, J. M., Edwards, K. R. & Poser, C. M. (1980b). Psychological profiles in patients with multiple sclerosis. A preliminary investigation. *Archives of Neurology, 37,* 437–440.

Poser, C. M. (1980). Exacerbation, activity and progression in multiple sclerosis. *Archives of Neurology, 37,* 471–474.

Pratt, T.T.C. (1951). Investigation of the psychiatric aspects of disseminated sclerosis. *Journal of Neurology, Neurosurgery, and Psychiatry, 14,* 326.

Shontz, F. (1955). MMPI responses in patients with multiple sclerosis. *Journal of Consulting Psychology, 19,* 74.

Sugar, C. & Nadell, R. (1943). Mental symptoms in multiple sclerosis. *Journal of Consulting Psychology, 98,* 267–280.

Surridge, D. (1969). An investigation into some psychiatric aspects of multiple sclerosis. *British Journal of Psychiatry, 115,* 749–764.

Weinstein, E. A. (1970). Behavioral aspects of multiple sclerosis. *Modern Treatment, 70*(5), 961–968.

Young, A. C., Saunders, J. & Ponsford, J. R. (1976). Mental change as an early feature of multiple sclerosis. *Journal of Neurology, Neurosurgery, and Psychiatry, 39,* 1008–1013.

# 9

# Psychological Problems Associated with Facial Disfigurement

*Norman R. Bernstein*

## PSYCHOLOGICAL ASPECTS OF STIGMA AND DEVIANCE

John Mortimer (1982, 91) wrote about

> the harsh inequalities caused by beauty. We could, if we had any real intention of doing so, narrow the wage differential, we could make education, spectacles, false teeth, and rides on the underground open to all, regardless of the accident of birth. No power on earth, however, can abolish the merciless class distinction between those who are physically desirable and the lonely, pallid, spotted, silent, unfancied majority. It is this class envy, I suppose, which makes men behave with a particular boorishness to the prettiest girls; it is what causes the sigh of relief when you have followed a back with admiration and then find a face which is unchallengingly plain. It is why the best news that an envious world often hears about a beautiful woman is that she has "gone off."

There is a border between plainness and disfigurement. Macgregor (1979) says disfigured people find "cold comfort" in the words of the Bible, "Man looketh on outward appearance and God looketh at the heart." The face is our passport to the world. Disfiguration often forces emigration from the familiar community and through this shift a loss of one's cultural roots. Yet, it is not always so.

President Eliot of Harvard University, a notable figure in education during his lifetime (1834–1926), was born with a cavernous hemangioma of the right side of his face. While this disfigurement caused him great discomfort and suffering, it did not stop him from pursuing a professional life and having a family. However, in his official portrait at the university, he is posed carefully in left full face so that his disfigurement is not visible. Artists in royal courts have always tried to eradicate imperfections in the countenances of their employers to finish the improvement that nature had "neglected" and thereby depict the "true perfection" of their models' souls. This is a case of our general policy to hide the defects of public figures. And it is so deeply imbedded in all of us that our leaders should be acceptable looking that it is difficult to think of any historical figure who was facially disfigured. In fact the worldwide dissemination of television has made facial appearance and expression so important that almost all politicians are specially clothed, made up, and posed to look their best. It is not that "the medium is the message," but certainly that the personal envelope of appearance shapes the message whether it is about the city water tax or nuclear defense. In relations between people, the medium of communication is to a great extent facial, and the appearance and attractiveness of the transmitting instrument is a large part of the impact of the shared information (Birdwhistel, 1970; Goffman, 1956; Goldwyn, 1986).

*Case 1.* Robert L. was a 26-year-old man who had been badly burned in an explosion at his factory. A year after his injury, he had red, raised scars and masking, fixation, and discoloration of the face. His nose was scarred, with only residual nostrils, which were red and angry looking. He bitterly described the loss of acceptable appearance, women friends, and companionship with his buddies, who were going on in their social lives and felt uncomfortable taking him along. He described the time he wore a compression garment (Jobst mask) on his face when he went to the supermarket only to have a man from a shop, thinking he was a robber, come up behind him and stick a shotgun in his back. His life was marked by social constriction, failure to return to work, and transformation into a chronically disabled person even though he had only modest loss of function in his hands.

*Case 2.* Rudolf S. was a 71-year-old man who had his left eye and upper face resected for cancer. He had a crater surrounding his eye, which was ultimately covered by a prosthetic device made at the university hospital. The prosthesis included an artificial eye and was held in place by eyeglasses. During the year between his surgery and his reception of the cosmetic device, he almost never left the house, except to come to the hospital, in spite of the urging of his wife and three grown children. He was partially retired at the time of the operation but withdrew from any work during the year of rehabilitation and the period of waiting for what he called his "passport" to the world, his cosmetic appliance. With this he could feel whole, and while close inspection revealed that it was not real flesh, it enabled him to resume social and work activities.

*Case 3.* Arlene R. was a single, 35-year-old bank executive. She had grown up with moderately severe acne, which did not subside until she was in her middle twenties, leaving her permanently disfigured. She always assumed that the modest pitting that was left in her skin was noticeable to everyone she met. It was a constant burden to go into a new situation where she felt her "defect" drew constant attention. She dated and received no outside reinforcement of her imagined unattractiveness, but it was imbedded in her self-concept that she was not appealing, and she telegraphed this attitude to people by her clothing, her shyness, and her apprehensive social bearing. This attitude overflowed into thinking that her work was not top drawer even when she did well and was promoted. There was a contagion of feeling from the visible facial defect to inner ones.

*Case 4.* In the book *Race for Life* (Sonnenberg, 1983), the mother of a badly disfigured child describes how he was injured in a multiple-car accident when their vehicle was struck full force by a 40-ton tractor trailer truck. Twenty-two–month–old Joel was trapped in his infant seat in a blaze of twisted metal and survived, suffering third-degree burns over 85% of his body. He lost all his fingers and toes, and his face became severely disfigured with masking of his features from scar tissue and distortion of his eyes, nose, and lips. His mother describes going into a shoe store to get Joel his first shoes after the accident and seeing another mother and her son staring at Joel.

> I knew the mother probably wondered what happened. I decided to be the first to speak. "Joel was burned in a car accident. That's why he looks the way he does." I felt stupid saying this but I felt forced to explain Joel—somehow needing this woman's positive or empathic response. Her response? She ignored me. Like I didn't exist. She moved a few chairs further over. She kept watching her son at the fish tank. This was worse than if she'd said "So what?" because at least then I could have yelled something back. Instead, her silence made Joel and me invisible, as if we didn't exist. I felt like I had been cut by a knife. (Sonnenberg, 1983, 148)

This is a sharp example of how such disfigurement intrudes in the development of the individual as well as the functioning of his or her family. For many nondisfigured people there is an additional reaction of *having to look,* of forcing oneself to see the horror, of simple satisfying curiosity, or of being reassured by the misfortune of another *(Schadenfreude).*

## SOCIOLOGICAL PERSPECTIVES

It is quite difficult to classify disfigurements, and we tend to discuss the many varieties in uniform terms. Nicola Rumsey (1983) rated disfigurements under the following categories:

**1.** *No disfigurement.* Your face is within the range of normal or usual appearance, and you do not feel it is damaged or malformed in any way that would justify the label disfigured.

**2.** *Mild disfigurement.* Your face is to some extent disfigured, but the disfigurement is neither conspicuous nor apt to attract attention. You will usually need to call attention to it before it is noticed by other people.

**3.** *Moderate disfigurement.* Your face is notably disfigured. People sometimes make remarks or question you about your face. Some people may tease you or stare at you, but in general there is no violent, negative reaction from other people.

**4.** *Marked disfigurement.* There is definite, notable facial disfiguration. Your face produces strong reactions from other people, such as repulsion, jokes, pity, curiosity, deliberate avoidance, or undue staring.

**5.** *Gross disfigurement.* Your face is definitely shocking and repelling to others. People react violently with horror, repulsion, pity, or the like.

There is no precise way of comparing disfigurements. Paralysis of the face may "disfigure" more than discoloration. Loss of an eye or absence of an ear may have a valence that is more striking than a flat scar. Scarring that cuts across the nose is more disturbing than scarring on the side of the nose alone. Many authors have described the facial triangle bounded by the eyes and the point of the chin, in which most significant disfigurements are detectable and produce the greatest social handicap (Freud, 1935/1952).

A disfigurement is not merely an individual deformity or an isolated event. It is a stigma that interferes with the interpersonal relations of the individual. While concepts of beauty vary greatly and the reactions to beauty cover a range from aesthetic contemplation to sensuous enjoyment and sexual desire, the continuum of reactions to disfigurement is largely negative, including the shock of recognition, puzzlement, apprehension, fear of physical contact or social interaction, horror, revulsion, and terrified avoidance.

There is a rare individual who is aroused or intrigued by disfigurement, but the majority have been acculturated to consider disfigurement as an affliction that is to be avoided. The presence of a facially disfigured person produces in others many of the reactions that chronic disease produces even though no actual disease may be present. One of these reactions is stress:

Stress is defined as a situation which places an adjustive demand on the individual. When situations are weighted to indicate the relative strength of their adjustive demands, the death of a spouse equals a hundred and Christmas equals twelve but these weights vary for each individual. Therefore, in assessing the stress of any one situation for any one individual, we must consider those factors which affect severity of stress: one, how long the stress lasts; two, the individual's tolerance; three, multiplicity of stressful situations; four, the individual's perception of the situation including whether stress is combined with threat.

Finally, if stress is combined with threat, the severity of the perceived stress will be greater than in a non-threatening situation. A flat tire on the way to an important appointment can create stress but the stress is compounded if the flat tire occurs in an undesirable neighborhood at night. (DeLoach & Greer, 1981, 20)

Any *disability* creates stress because it places many demands for adjustment on individuals. The sighted reader can at least partially understand the adjustment demands placed on an individual who loses his or her sight. Major facial disfigurement additionally destroys the place of the individual in the fabric of society, for the (usually permanently) disfigured individual is feared and avoided by others. This has been described as the leper phenomenon (Bernstein, 1976). Thus the major source of stress for the facially disfigured person comes from the environment, that is, from the reactions of others. Psychoanalysts have described it as a castration threat. According to Rumsey (1983) and many other social scientists, people react to disfigurement with consternation because they do not know how to react. The observer's discomfort with and avoidance of the facially disfigured person is a strikingly strong phenomenon that every professional who works with such an individual experiences and must come to terms with.

## UNDERSTANDING AND RELATIONSHIPS

There is a question of comprehending what has happened when a person is disfigured. Is the disfigurement due to neurological atrophy? Is it the acne or protruding teeth that trouble the adolescent? Is it ablation due to cancer surgery or trauma from an automobile accident or a fire? Helen Keller (1957, 5) said, "Not blindness, but the attitude of the seeing to the blind is the hardest burden to bear." Ashley Montague (1978) makes the point that the skin is the largest organ of the body and that making physical contact through touching is one of the most basic human experiences. Yet, in the presence of facially disfigured individuals, onlookers have a horror of touching and feel more comfortable as they get farther away from the disfigured individual until they are finally out of view. Goffman (1967) points out that every person is a stimulus to other people and that class affects facial expression, producing deference on one side and haughtiness sometimes on the other. He writes:

Ugliness then is a stigma that is focused in social situations. Other stigmas such as a diabetic condition are felt to have no initial effect on an individual's qualifications for a face-to-face interaction. They lead us first to discriminate in such matters as job allocation and affect immediate social interaction only for example because the stigmatized individual may have attempted to keep his differentness a secret and feels unsure about being able to do so. or because the others present know about his condition and are making a painful effort not to allude to it. There is a popular notion that although any contacts between strangers are particularly subject to stereotypical responses, as persons come to be on closer personal terms with each other, this categorical approach recedes and gradually sympathy, understanding, and a realistic assessment of personal qualities takes its place. While a blemish such as a facial disfigurement may put off a stranger, intimates, presumably, would not be put off by such matters. The area of stigma management, then, might be seen as something that pertains mainly to public life, to contact between strangers or mere acquaintances, to one end of a continuum whose other pole is intimacy. (Goffman, 1967, 50)

This is patently not the whole story. If public associations are restricted, then disfigured persons are cut off from many of the chances to become intimates. Americans are notably intolerant of being close to friends who have been disfigured and tend to become remote from them. In *Freaks,* Leslie Fiedler (1978, 13) notes the many types of disfigurement in society: dwarfs, giants, fat ladies, living skeletons, bearded ladies, feral children, hermaphrodites, elephant men, Siamese twins. He views them as society's badge of shame, a reminder of the long exclusion and exploitation by other humans, who by defining them have defined themselves as "normal." He mentions Freud's idea of seeing forbidden things when we look at these "freaks." P.T. Barnum called them curiosities, while others called them prodigies or monsters. Seeing disfigurement is dangerous and is a key to primitive thoughts, a threat to our own integrity and normality.

There is an old tradition of physiognomy (reading faces) going back to John Casper Lavater, a Swiss minister in the 18th century. He attempted to assign characterological traits to particular shapes of the nose, lips, and other features, because "the moral life of man particularly reveals itself in the lines, marks, and transitions of the countenance" (Lavater, 1783, 9). This is an ancient idea and is very well engrained in the behavior of modern individuals who, though they disavow the belief that people who look pretty are better, act as if attractive people are in fact more valuable than ugly people. This runs through our society: the handling of children with cleft lips—often regarded as a sign of mental defect—is an example.

There is considerable psychological research on the development of emotions that relate to facial appearances. Stephen Suomi (1984) describes research on the expressions of rhesus monkeys, which show fear, grimaces, and other nonverbal behaviors that are an innate, maturational-dependent response system, which rapidly becomes subject to environmental contingencies as development progresses. He compares this to the emergence of the fear of strangers studied in human infants in many different cultures. The question persists whether the fear of disfigured people is innate or acquired. Some of our visual sets appear very early. Newborns show "innate" smiles and expressions and early orientation to the face in primates. However, as infant psychiatrists have shown, babies learn very early to look at their mothers' faces even when nursing, and they respond to their mothers in a variety of ways, which convert intrinsic smiling into a learned pattern of interaction with the mother. Motion of the face, the appearance of the eyes, and hairline seem to be part of the earliest signals that children respond to, and these are homologous with infant behavior in other primates. Berscheid et al. (1972), in a series of publications, have dramatically delineated the ways in which attractiveness influences how people are considered and valued.

## THE PERSON BEHIND THE FACIAL MASK

Individuals who have undergone facial disfiguration will sometimes remark that they are the same person behind this disfigurement (Macgregor, 1979). However, they show marked alterations in their self-concept, their interpersonal style, their hopes for the future, and other qualities, which we put under the heading of normality. Lifton (1967) noted in the burned survivors of Hiroshima that there was a psychic "closing off" to suppress the survivors' rage or, in a broader sense, a resistance toward the forces manipulating them. This is related in part to the duration of their problems (Cassileth, et al., 1984). It takes years to readjust one's sense of identity and self-concept to manage a major disfigurement. Under the weight of enormous, continuous stress, hopefulness about the future is diminished and autonomous action becomes a constant challenge. The ability to empathize with others is frequently diminished because facially disfigured persons have to use all of their energies to cope with their own problems. Similarly, morality and humor suffer, and mastery of the environment, the capacity to love others, and the ability to perform work are all made more difficult by the burden of facial disfigurement (Bernstein et al., 1988).

## The Self-Concept

Rogers (1951) says that the self-concept of individuals who are becoming fully functioning persons will show that they are moving toward being themselves, toward meeting their own expectations rather than those of others, toward greater acceptance of others and of self, toward guiding their own lives, and toward openness in self-experience. It is clear that these formulations shift in the deformed person because society will simply not permit the facially deformed individual to maintain relatively normal, self-conceptual attitudes. The self-concept (Burns, 1979; Jeppson, 1963; Schilder, 1950) is largely described in terms psychiatrists would put under the heading "ego." Allport (1955) attempted to evade the confusing applications of self-terminology by using the term "proprium." The proprium consists of seven aspects:

1. Bodily sense
2. Self-identity through time or continuous existence
3. Self-enhancement and love of self
4. Self-extension identification with others and other things
5. Rationality, planning, and coping
6. Self-image
7. Appropriate striving, motivation, and behavior to enhance the self-image

While the self is the center of our lives and while some facially disfigured people remark that their selves are the same in spite of the disfigurement, it seems evident that the intense pressure upon them to conform to the outside world must alter their self-concept. Clearly this relates to what Schilder (1950) described as the body image—the shifting picture that we have of our bodies and ourselves—which changes in social situations but is determined by physical development, experience, and the environmental matrix in which we find ourselves.

William James (1950, 1:294) said, "A man has as many social selves as there are individuals who recognize him and carry an image of him in their mind." There is also a discrepancy between what people feel they are and what they imagine other people think about them.

## Development of the Social Self

At birth a child who has a cleft palate, a strawberry mark, or a large cavernous hemangioma produces a shock in his or her mother and an alteration in the earliest stage of dependency on the mother. Her reactions shift their interactions. Once there is a shift in the line of development, it produces compensatory reactions in the family. The child gets worried over more and receives more attention; sibings react to this in a variety of ways. As the child begins to walk and to play with other children, these playmates begin to comment, remark, and ask questions about the defect. School-age children often seem to handle these remarks and jokes fairly well, even though such comments engender a sense of inner discomfort and insecurity. Parental attitudes of rage toward the public and protectiveness may further skew the pathway of development.

By the time adolescence is reached, with its many alterations and the intense social challenges and experimentation that occur, the facially disfigured child feels more sharply thrust to the side, into a deviant role. This child is less able to compete in all the social activities and role experimentation that are part of the vigorous and stressful psychosocial period between latency and adult maturity. At this age the sense of identity becomes crystallized. Yet the child with a disfigurement may feel less competent or less attractive. While the published stories tend to be tales of triumph, for patients with cleft palate the psychological studies show a chronic struggle against feelings of inadequacy and stigma. For some young men the ability to grow a moustache and cover up the cleft is an enormous progressive leap, and some will then put disfigurement largely behind them. Analysts have described many fantasies that accompany the intrapsychic adaptation to physical handicap. There is grandiose overcompensation in some of the dreams of glory, but these are not usually recounted directly by patients.

Individuals with extreme disfigurement undergo what Frances Cook Macgregor (Macgregor et al., 1953) described many years ago as "social death," where they withdraw from all social relations and stay home as "closet people," who have dropped out of the ordinary interactions of the community and disappeared from the world of normal relationships.

## INTERVENTIONS WITH DISFIGURED PERSONS

Interventions to help facially disfigured people are fundamentally different from other forms of psychotherapy or rehabilitative work. First, treatment is a battle against societal values and community reactions. Second, it involves a problem of appearance and symbolism rather than a problem of mobility, pain, or sensory loss. And finally, there is an element of chronic grief over the situation. It is critical to be aware of the time scale: for most of these individuals adjustment to disfigurement takes years and rehabilitation takes years more—sometimes a lifetime.

## Early Stages of Intervention

In the early phase of intervention, a medical model is often used. Patients are depressed, they experience severe pain, and they have to come to grips with their situation. The hospital team is in charge, and the nurse is the keystone of this team. She is involved as a planner, a provider and coordinator of care, and a patient educator. Occupational therapists, physical therapists, teachers for school-age children, psychiatrists, social workers, psychologists, ward aides, and attendants all help and support the patient. The main thrust is toward activity, getting the patients to do more for themselves, to cope with their problems, to confront unpleasant procedures, and to understand all procedures so that hope is maintained along with some realistic tough-mindedness. The team members can share with each other what they have told the patient, what the patient has or has not inquired about, such as problems with eating, thoughts about ever having sexual relations again, kissing his or her children, and so on.

Early family involvement is important. Family members need to go through a mourning and bereavement process, which involves giving up the old image of the person and accepting the new one, with altered goals and new limitations.

During this time, patients frequently have sleeping and eating disturbances and become apathetic and dejected. Antidepressants have been useful in some cases, but the major helping instrument at this time is caring people. Activation is crucial; allowing visitors, encouraging attendance in groups, involving the family in groups, and giving patients free use of telephones are helpful.

Vocational planning should begin early. It should be proactive rather than reactive. The sooner the patient can contemplate part-time work, light work, or limited work, the better. An emphasis on mastery is important with these patients so they can look ahead to getting up and around and to restoring their previous competence. Patients need support and preparation in going from hospital to home and ultimately from home to school, to work, and social situations. They may go back and forth several times before they finally take the decisive step toward reentry into social life.

## Litigation

For people involved in accidents, there is an increasing emphasis on suing for compensation. Since America is the most litigious society in the world, this is not at all surprising. It is striking, however, that involvement in litigation has a major supportive effect on patients. It is clear that when people can externalize their anguish by blaming others and when they can focus their rage and hurt with help from enthusiastic lawyers, these situations frequently improve mood, focus tension, and give an immediate purpose to their lives.

## Later Phases of Intervention

The long-term phase was defined by Cromes (1984, 345) as "the interval from hospital discharge until all treatment programs are completed and the patient has settled into a satisfactory living pattern." This is both correct, in one sense, and too optimistic in another, since it assumes that all patients will enter a satisfactory living pattern. In reality we know that a number of facially disfigured people live in seclusion and do not have an existence that is at all gratifying. We have no good estimate for the number of people who go beyond social death to actual suicide. Many disappear from sight.

Our goals are to restore patients to as much normality as is feasible. A major feature of intervention with disfigured people is the use of cosmetics and cosmetic surgery. This is a response to the realities of life in our society, that the better these patients look, the better their adjustment will be.

Minor disfigurements will sometimes be as psychologically disastrous as major ones. This has to do with the multiple images people have of themselves and also with the fact that people with minor disfigurements attempt to have an active social life, thus exposing themselves to a large number of rejections, while people with major ones often retreat to another level of social interaction, which removes them from unpleasant conflict and exposure. Janet Jeppson (1963) did not observe a signficant correlation between the degree of psychological improvement and whether the patient was satisfied or dissatisfied with the surgical results. This fact should be taken into consideration by anyone working with such patients. If a psychiatrist, for instance, has a patient who undergoes plastic surgery, the fact that the patient is pleased with the results is no guarantee that any psychological change for the better has taken place.

## Job Placement

Janet Pinner (1953, 154) states:

Regardless of what some people might believe, job placement of the handicapped is not done through the use of snappy slogans. These slogans are the product of public relations men who try to make the community aware of what the handicapped can do. The publicists have done an excellent job. But the public's image of the disabled is not that of a facially disabled person. Some employers are convinced that if you hire the handicapped, you will have less turnover, better efficiency and less absenteeism. If one were to believe all the publicity, no one would hire anyone but a handicapped person. Professionals know better. It's individual differences that really count. There must be sufficient time for the placement specialist to do an individualized job. What kinds of jobs do people with facial disfigurement obtain? This is very interesting: the placements range from a clerical level to the unskilled, with three-quarters of the placements being made in service in unskilled occupations. We have the feeling that many of these workers go into service occupations because these are the jobs behind the scenes for unskilled workers. These are the jobs where the employee works for himself, away from the general public, and therefore cannot be seen.

Employment must be individualized and aggressively pursued. The patient must realize that repeated failures will occur and that the longer one is out of work the harder it will be to get back to regular employment.

## SEXUAL FUNCTIONING

Cromes (1984) reports that for men loss of libido is a more common complaint than for women, while in women orgasmic dysfunction seems to be more of a problem. Both of these problems are clearly related to the presentation of self. These persons' views of themselves and their self-concepts have been shifted, their relationships with sexual partners have been altered, and their relations between themselves and social values have also been changed. Family members need support and instruction in dealing with these changes, and they also need an opportunity to vent their own feelings, their shock, and their own needs for satisfactory sexual stimulation. These changes constitute one more handicap to self-esteem and self-actualization for facially disfigured individuals. Graceful and concerned inquiry by the health care worker is the key here.

## STYLES OF FUNCTIONING

To be out in public requires emotional work when the public does not accept one's appearance. Individual techniques for this challenge include the psychiatric devices described by Bibring and Kahana (1954), who include support, education, clarification, confrontation, and interpretation as well as suggestion and manipulation. Patients need to abreact and vent some of their own feelings; they need to assimilate what has happened to them and adapt themselves well. Kahana and Bibring point out that manipulation is a vital part of care in any psychotherapy and a major part of the activities of therapists working with facially disfigured persons. This includes pushing patients to go out, calling schools or jobs, getting cosmetics for them, helping them with prostheses, and changing the scheduling of surgical interventions. Support is obviously important and should come from the network of supporting people. The family in particular needs to be pursued; frequently, evaluation in the home is necessary.

The health care professional must be in touch with these patients aggressively in order to get optimal results with rehabilitation. Some groups of patients are notoriously unresponsive. Veteran hospitals report that many of their ablation patients, who have had disfiguring surgery for head and neck cancer, are not cooperative. This is a particularly troublesome group because of the high incidence of smoking and drinking and the way these patients become devalued by staff, a response that the patients perceive strongly and one that contributes to their apathy and withdrawal. Apathy is one of the many depressive feelings in facially disfigured patients. Sometimes antidepressants can help, but most helpful still is an aggressive, total-push program to get the patients going in the world.

Though one cannot be certain about outcome, one should always aim higher in order to enhance the patient's functioning. To give up too quickly seems to condemn the patient to an emotional collapse and the likelihood of reclusiveness and constriction of life. The push should involve family; it should involve ongoing surgical planning, because nothing helps psychologically as much as the hope that future surgical intervention will lead to further improvement, even when this improvement turns out to be minor (Macgregor, 1982). There is a lesson to be learned from the women's movement with its emphasis on networks. Women's groups have repeatedly stressed that networking helps support them with the difficulties and discrimination they face on the job. Networks of disfigured people can be very helpful and are not used nearly as much as they should be. Individuals show vast differences in their abilities to reach out.

In World War II, injured Royal Air Force pilots who belonged to the "Guinea Pig Club" were treated by the same doctor. They wore a blazer and formed an organization that has continued to meet for 40 years. This shows how solidarity within a group helps maintain personal value and how it aids the coping process.

Patients need to be pursued. Underlining what the surgeon is doing is often the most useful way to keep in touch with them. Focusing on cosmetics and prosthetic appliances can help alleviate their hopelessness and giving-up attitudes. Professionals have to mix a certain energetic determination with realistic flexibility about when to give the patients time to rest and when to push and activate them. Although we know that many facially disfigured patients will not respond to the best, most energetic treatment, we are fully aware that without an enduring, all-out effort, many more will be lost to useful social lives.

## REFERENCES

Allport, G.A. (1955). *Becoming.* New Haven: Yale University Press.

Bernstein, N.R. (1976). *Emotional care of the facially burned and disfigured.* Boston: Little, Brown.

Bernstein, N.R., Breslau, J.J. & Graham, J.A. (Eds.). (1988). *Coping strategies for burn survivors and their families.* New York: Praeger.

Berscheid, E., Walter, E. & Bohrnsted, G. (1972). Beauty and the best. *Psychology Today,* 5(10), 42.

Bibring, E. & Kahana, R. (1954). The dynamic psychotherapies. *Journal of the American Psychoanalytic Association,* 20, 746–769.

Birdwhistel, R. (1970). *Kinesics and context.* Philadelphia: University of Pennsylvania Press.

Burns, R.B. (1979). *The self-concept: Theory, measurement, development, and behavior.* London: Longman.

Cassileth, B., Lusk, E., Strouse, T., Miller, D., Brown, L., Cross, P. & Tennglia, A. (1984). Psychosocial status in chronic illness. *New England Journal of Medicine,* 311(8), 506–511.

Cromes, G.F. (1984). Psychosocial aspects. in S.W. Fisher & P.A. Helms (Eds.), *Comprehensive rehabilitation for burns.* Baltimore: Williams & Wilkins.

DeLoach, C. & Greer, B.G. (1981). *Adjustment to severe physical disability.* New York: McGraw-Hill.

Fiedler, L. (1978). *Freaks: Myths and images of the secret self.* New York: Simon & Schuster.

Freud, S. (1952). *Civilization and its discontents.* In J. Strachey (Ed. and Trans.), *The standard edition of the complete psychological works of Sigmund Freud* (Vol. 21). London: Hogarth. (Original work published 1935.)

Goffman, E. (1956). *Stigma: Notes on*

*the management of spoiled identity.* Englewood Cliffs, N.J.: Prentice-Hall.

Goffman, E. (1967). *Interaction ritual.* New York: Anchor.

Goldwyn, R.M. (1986). *Beyond Appearance: Reflections of a plastic surgeon.* New York: Dodd.

James, W. (1950). *The principles of psychology.* New York: Dover.

Jeppson, J. (1963). Psychiatric problems in the facially disfigured. In B.O. Rogers (Ed.), *Facial disfigurement: A rehabilitation problem.* Washington, D.C.: Vocational Rehabilitation Administration.

Keller, H. (1957). *War blind in American social structure.* New York: American Foundation for the Blind.

Lavater, J.C. (1783). *Essay on physiognomy* (15th ed.). London: Wm. Tegg.

Lifton, R. (1967). *Death in life: Survivors of Hiroshima.* New York: Random House.

Macgregor, F.C. (1979). *After plastic surgery: Adaptation and adjustment.* New York: Praeger.

Macgregor, F.C. (1982). Symposium on social and psychological considerations in plastic surgery. *Clinics in Plastic Surgery, 9,* 3.

Macgregor, F.C., Abel, J.M., Bryt, A., Laver, E. & Weissmann, S. (1953). *Facial deformities and plastic surgery: A psychosocial study.* Springfield, Ill.: C.C. Thomas.

Montague, A. (1978). *Touching: The human significance of the skin.* New York: Harper & Row.

Mortimer, J. (1982). *Clinging to the wreckage.* London: Penguin Books.

Pinner, J. (1953). Vocational placement of the facially disfigured. In F.C. Macgregor (Ed.), *Facial deformities and plastic surgery: A psychosocial study.* Springfield, Ill.: C.C. Thomas, 154.

Rogers, C. (1951). *Client-centered therapy.* Boston: Houghton Mifflin.

Rumsey, N. (1983). *Psychosocial problems associated with facial disfigurement.* Unpublished manuscript.

Schilder, P. (1950). *The image and appearance of the human body.* New York: International Universities Press.

Sonnenberg, J. (1983). *Race for life.* Grand Rapids, Mich.: Zondervan.

Suomi, S. (1984). The development of affect in rhesus monkeys. In N.A. Fox & R.J. Davidson (Eds.), *The psychobiology of affective development.* Hillsdale, N.J.: Lawrence Erlbaum.

# PART IV
## Social Psychological Issues

The chapters in part 4 deal with social dimensions of physical disability not specifically covered in the preceding chapters. These include the psychosocial impact of assistive devices for persons with disabilities, sexual concerns and their treatment, forensic psychiatry and disability, ethical and legal issues, and the independent living movement.

The impact of technological advances on improving the functioning of persons with disabilities is a major development of the past 20 years. Jean Kohn describes the psychosocial effects of the successful integration of assistive devices in the lives of disabled persons and offers a model in which the combination of the person and an assistive device interacts more successfully with the environment. She begins with a brief review of the factors that have promoted the striking number and variety of assistive devices developed in the past two decades and the benefits that have been potentiated by these devices. For young children, such devices can facilitate normal developmental patterns and minimize potential secondary emotional problems. For adults, assistive devices can replace lost capabilities or maximize functioning impaired since birth. For disabled persons across the age span, assistive devices can provide structural support to increase comfort, prevent progression of deformity, and facilitate physical management and independent living. Kohn discusses how the right "match" of assistive device, disability, and personality can prevent or alleviate stresses associated with disability, particularly difficulties associated with feelings of loss of control. She underlines the importance of consumers choosing specific devices for themselves and the necessity for consumer input in the future development of assistive devices.

Susan Knight discusses sexual problems in persons with disabilities and suggests interventions for their treatment and rehabilitation. Given the wide range of disabilities as well as other significant variables such as present age, age at onset of disability, previous sexual awareness and experience, and family and peer support, it is difficult to specify step-by-step interventions for every sexual difficulty. But most commonly, clients experience sexual discrimination and body-image concerns in addition to more specific difficulties.

At the outset, these two concerns must be dealt with in an empathic and focused manner. The sexual attitudes, knowledge, and experience of individuals (as well as those of their sexual partners and family members) must be explored. At the most basic level, clients may lack information or may have attitudes about sexuality that hinder their satisfactory adjustment. Sexual adjustment for congenitally disabled persons is often difficult because, unlike able-bodied children, who "take" their independence, disabled persons often must be "given" it. They often strongly internalize their parents' negative attitudes about sex and disability. At this level, the therapist must help foster a sense of identity, which includes sexuality as a permissible, expectable, and even highly pleasurable activity. For clients who lack sexual experience, sexual surrogates can be useful, enabling clients to experience what sex with a partner is like as well as to discover what, if any, part their disabilities play in sexual expression. Counseling focused on assertiveness, and the exploration of sexuality through masturbation may be useful. Sexual feelings and responses need not reside only in the sexual organs or occur only via the usual means of stimulation. In addition, common but false stereotypes—"wheelchair-bound persons do not experience genital sensation and sexual response," for example—must be addressed.

The sexual concerns of the partner of a disabled person must also be dealt with, especially in cases of traumatically acquired disability; counseling for couples may be indicated. Finally, social skills training and group therapy are often helpful. If a disabled person can respond to stereotyped reactions with a lack of defensiveness and in unexpected ways that promote comfort on the part of the other, and if he or she can communicate a solid sense of self and a lack of feeling victimized, then more satisfying and spontaneous interactions may result.

A key feature of disability is the interface between disability and work. Carroll Brodsky's chapter focuses on the methods and concerns of the forensic psychiatrist in evaluating cases of disputed disability claims in which physical and psychiatric impairments overlap. Drawing upon his training in anthropology as well as psychiatry, Brodsky delineates the vocabularies, values, rules, roles, practices, beliefs, and ethics that differentiate clinical from forensic psychiatric practice. He depicts the often-strange "looking glass world" of forensic psychiatry, where the "no fault" system for providing benefits to injured workers has evolved into an arena for lawyers, physicians, and other specialists in which medicolegal costs add substantially to the system. He details how the worker's compensation system works, where claims may be decided, which issues may be disputed, why disputes are raised, and the bases on which they may be settled.

Brodsky describes the different approaches of the forensic and clinical psychiatrist to their relationships with the client or patient, to questions of causation, to the determination of a disorder's presence, and to its epistemology. Depending on the issue and the forum involved, the forensic psychiatrist evaluates the pain and suffering, mental distress, and/or the

# INTRODUCTION

psychiatric disability associated with a physical or functional disability. Issues of monetary compensation and quantification of disability are almost always involved. Brodsky describes the methods of the forensic psychiatrist, the assignment of psychiatric diagnoses, and particular challenges—in giving testimony, in the quantification of distress, impairment, or injury for assessing suitability for work, and to professionals' integrity—faced by practitioners in the field. Finally, he discusses the impact of litigation on the disabled claimant and on postlitigation adjustment.

Robert Nelson discusses the tangled ethical, legal, and social issues related to disability from the dual perspective of being both an ethicist and a specialist in pediatric critical care. He begins by describing the recent controversy surrounding the application of Section 504 of the Rehabilitation Act of 1973 to the "Baby Doe" case, in which the United States Supreme Court held that Section 504 did not apply to treatment decisions in which the disability itself is a factor in determining benefit from treatment. In these cases, the Court held, Section 504 should be set aside. The central issue, as Nelson sees it, is "whether the presence and extent of a physical impairment is sufficient reason for not treating an infant who otherwise would be treated." And this issue, he argues, is just one instance of a broader theme: the status of disabled persons within our society. Does the Court's recent decision imply that a disabled person may be denied employment if the disability undercuts efficiency? Or should it be interpreted in a more circumscribed manner—that Section 504 should not be considered in individual medical decisions? In this sense, such issues as justice versus legality, and ethics versus cost-benefit ratios assert themselves in apparent opposition.

There seems to be an inherent tension, Nelson suggests, in the proposition that individual rights are based on equality. Presently, persons with disabilities are accorded individual legal equality and offered full participation only by ignoring the disability, while at the same time the case is made for special accommodations on the basis of disability. Nelson holds that the ethical relevance of disability must be established in order to provide a basis for social entitlement and that the context or purposes for which the judgment of equality is made must be specified. For example, the question of disability may be irrelevant to considerations related to education and employment, while it may be quite relevant when considering medical treatment. The question then becomes, what differences among persons make a difference in terms of specific goals, obstacles, accommodations, and benefits? Ultimately, Nelson suggests, the demands of justice must be tempered with the principle of compassion.

Edward Roberts, founder of the independent living movement and the World Institute of Disability, former director of the California State Department of Rehabilitation, and MacArthur Award recipient, has been severely disabled by polio for many years. His view differs from that of Nelson in that he regards the full participation of persons with disabilities as a right that must be guaranteed by the legal system rather than by distinguishing between "justice" and "compassion." He characterizes the most pervasive barriers to full participation by persons with disabilities as attitudinal rather than physical. The medical model, in his view, has severely hindered disabled persons in this respect. Because disability cannot be cured, because there is no "infectious agent," and because of the long-term nature of the work, professionals in rehabilitation are often accorded lower status and more meager financial rewards than their colleagues in other areas of health care. The irony is that it is precisely because of medicine's successes that much of the problem exists. Most people do not believe they will ever become disabled. By increasing our awareness that we are all temporarily able-bodied, the we-they dichotomy may be eliminated and the successful and productive integration of disabled persons into society can be seen as a benefit to all.

Disabled people have experienced a long history of discrimination. In the social Darwinism of the 19th and early 20th centuries, disabled persons were considered "waste," impediments to increased productivity. Institutionalization was common, and the removal of disabled persons from public view, while supporting the idea that they were sick, infected, and dying, was consistent with contemporary sensibilities. In the 1940s, the model was "complete recovery," and while the period saw enormous improvements in the status of disabled persons, the attitude that anything less than a complete return to "normal" functioning was a failure still maintained, to the sentiment of persons with disabilities, an able-bodied versus disabled dichotomy. It was also in the 1940s, however, that consumer groups began to organize, at first around such major issues as attendant care and home support. In the 1960s, the civil rights movement for members of ethnic minorities and integration into the mainstream became prominent. This current was soon embraced by consumer advocates for disabled citizens as well giving rise in the 1970s to the birth of the independent living movement.

The independent living movement employed advocacy and political action to promote equal rights for disabled persons. Independent living centers began to spring up all over the country. In addition to consumer advocacy and political action, these centers offered direct services, counseling, and assistive devices to disabled persons. Founded on a self-help model, independent living has intrinsic advantages when compared with the SSI/welfare model pursued by the government. Besides increasing productivity and enhancing quality of life, self-esteem, and psychosocial functioning to a greater extent than the welfare model, it is also, suggests Roberts, less expensive. The challenge, he asserts, is to establish enough independent living programs so that they are accessible to all who need them.

Albert Jonsen begins his chapter by describing the personal courage of a young woman with an exceptionally severe case of neurofibromatosis to illustrate the discrimination faced by persons with disabilities and the personal resources needed to cope with such a condition in a society that provides little or no support for such individuals. He discusses the derivation of the word *discrimination,* among others, to show how the current, value-laden connotation of this word has strayed far from its original, neutral meaning. He suggests that human beings create despised classes because of envy and jealousy, and he delineates six features of the social perception of disabled persons that account for their despised status and hinder their full participation in society. He closes by noting again the enormous courage shown and needed by persons with disabilities, given the toxic attitudes toward disability in our society.

# 10
# Psychosocial Aspects of Assistive Devices for Disabled Persons

## Jean G. Kohn

Medicine in general tends to look at pathology, at what is *wrong* with the patient. . . . Rehabilitation is unlike other areas of medicine in that it deals with patients whose pathology is generally permanent or irreversible. . . . Rehabilitation is concerned with what's *right* with the patient, with residual abilities, resources and skills. (Corcoran, 1982)

The process of rehabilitation enables disabled individuals to learn how their residual assets can be used in functionally adapting to their living requirements. Adaptation may range from passive acceptance of one's dependence on others in most areas of living to successful, independent participation in work and social activities. Disabled persons may require assistive equipment to reach levels of function compatible with their desires, and rehabilitation technology has developed in response to these expressed needs. Rehabilitation technology can, in fact, reduce the handicapping effects of disability (Vash, 1983). Appropriate use of such technology has been increasingly evident in the past decade and parallels the increased visibility of disabled people in the mainstream of American life.

The point of this chapter is that the combination of a disabled person and an assistive device results in a new, better-functioning entity. The functional improvement of a disabled person equipped with an appropriate device is manifested in greater independence, increased control of interaction with the outer world, and therefore, a sense of greater competence, integrity, and satisfaction in life. This is diagrammed in figure 10.1. If interac-

```
                    interface              (successful)
        PERSON              DEVICE ⇐Interaction⇒ ENVIRONMENT
                                        (unsuccessful)
```

PERSON includes physical function, self-perception, disability (static or progressive) and, if a child, developmental level

INTERFACE (or control) is the means of accessing the device (electrical control, seating system, etc.)

DEVICE may be a wheelchair, communication system, environmental control system, etc.

ENVIRONMENT may include people, physical settings, travel, work, recreation, etc.

**FIG. 10.1.**
Interaction of the Combination of Person and Device with the Environment

tions are successful, psychosocial effects are positive for both disabled persons and the people who interact with them. Positive interactions lead to an improved self-concept and to a validation of the disabled person's competence.

## BACKGROUND

During the past 20 years, many technological, legal, environmental, and social changes have resulted in increased use of assistive devices. There has been a marked change in the variety and construction of assistive devices for disabled people. The major changes in rehabilitation technology have paralleled the changes in technology as a whole. The development of lightweight, strong, synthetic materials and their methods of construction are based on similar developments of synthetic materials for the general public. Wheelchair development and changes in construction to meet the needs of recreational-wheelchair users have drawn from materials used in automobiles, strollers, and other equipment for the nondisabled population (Clearfield, 1976). Electronic devices, miniaturization of components, and the rapid changes of the computer revolution, all of which are used in rehabilitation technology, have drawn both from inventions directed toward business use and from space research (Office of Technology Assessment, 1982).

Legislative changes, which have resulted in the inclusion of handicapped children and adults in mainstream society, have encouraged the use of assistive devices to facilitate transitions into schools and communities.[1] Environmental changes, including "curb cuts," building and transportation alterations for improved access, and special parking places for disabled people, have also helped to encourage the use of assistive devices (Architectural and Transportation Barriers Compliance Board, 1977). But possibly the greatest factor in societal change for the disabled population has been its self-advocacy, spearheaded by spinal cord–injured veterans, postpolio paralytic adults, and the well-organized national groups of the deaf and the blind (White House Conference on Handicapped Individuals, 1978).

Advocacy and the increasing inclusion of disabled people in vocational and recreational activities in the same settings as nondisabled people have also prompted demand for assistive devices that will increase functional levels (Enders, 1984). Independent living programs are a result of the desire to live in the community even though requiring attendant care, adaptations of the home, and assistive equipment (Lifchez & Winslow, 1979).

Finally, the area of self-advocacy that has been unexpectedly successful is competitive sports participation by disabled individuals. This interest has led to a demand for durable, reliable, appropriately engineered equipment and has had an especially stimulating effect on the wheelchair industry (Fox, 1984). Disabled athletes have participated in the development of sports equipment for skiing, sledding, archery, and such wheelchair sports as basketball and track.

Concurrent with expressed needs for improved equipment, there has been a demand for information about what devices are available, where to get them, and how much they will cost. There is currently available a variety of national information-retrieval services (NARIC, ABLEDATA), which list thousands of items of assistive equipment (Enders, 1984). However, information is still inadequate; continued improvement in quantity and availability of such information to consumers is essential.

At present, the negative factors in the provision of assistive equipment are the service delivery and reimbursement systems; this combination limits access to appropriate devices for many disabled people. Several quotations from the Office of Technology Assessment monograph "Technology and Handicapped People" (1982) speak to this issue.

Services and funding for disabled people come from so many different, often uncoordinated sources that both providers and users often are unable to take advantage of available technologies. (133)

Decisions regarding who should be eligible and how eligibility should be determined are major determinants of the use of technologies. (132)

Despite eligibility for the public and nonpublic programs that may pay for technologies to assist them to function more independently, a number of disabled people are denied funding for particular technologies that are clearly appropriate. . . . A primary reason for the denial of funding is that the technologies in question are not strictly "medical" in nature and are therefore not considered "necessary." A connection needs to be made in these programs between paying for these technologies and the potential independence or productivity of disabled people. (177)

## PRESENT TRENDS

The practical advantages technology offers to disabled people can be grouped into three areas: developmental assistance, replacement assistance, and support or comfort for disabled persons and their caretakers.

### Developmental Assistance

For children, physical, occupational, and speech therapies are aimed at improving function through exercising, strengthening muscle power, and inhibiting undesired reflex patterns. Some children will never be able to carry out activities such as walking, talking, or using their hands successfully in activities of daily living. Yet these children, like all others, face the usual tasks of growing up: exploring one's surroundings, separating from parents into neighborhood and school settings, and interacting with peers.

Assistive devices can foster normal developmental patterns and reduce secondary emotional problems, isolation, and dependency. Their use can also promote cause-and-effect learning, the manipulation of objects to understand their characteristics and uses for both play and psychological expression (Tobias, 1983). Assistive devices may also enable the child to discharge feelings of anger, rebellion, or displeasure. This may reduce such negative behaviors as head banging and temper tantrums. Tobias observed, "Severely disabled children do not get much opportunity to 'slam the screen door'" (343). Assistive devices may provide the child with similar, acceptable substitutes.

To achieve greatest benefit, however, the devices must coincide with the child's level of psychological development as well as the physical disability. Sutcliffe (1972, 296) lists, as one aspect of care for handicapped children, "realism in choice and timing of remediation strategies to secure psychosocial growth." For example, children with cerebral palsy who are nonverbal will require tools to facilitate language development appropriate to their chronologic cognitively equivalent age (table 10.1). Similarly, those children who cannot walk independently require a sequence of assistive devices, which can provide the opportunity for mobility and appropriate interaction with the environment (table 10.2).

It is not necessary to provide high technology devices in all situations; in fact, younger children may find simpler devices more suitable to their de-

**TABLE 10.1**

Developmental Aspects of Normal Verbal and Written Communication and Alternatives for the Nonverbal Child

| Age | Verbal child | Nonverbal child |
| --- | --- | --- |
| 8–12 months | Words: yes, no | Nodding, body movement, facial expression |
| 12–24 months | Single words, two-word sentences | Picture board (eye movement, finger and hand pointing) |
| 2–5 years | Sentences: How? Why? When? I want . . . I feel . . . | Blissymbolics;[a] simplified computer use with pictures |
| 5–10 years | School tasks: reading, writing, speaking in class, etc. | Computer with adapted access and voice output |

Source: Compiled from Musselwhite & St. Louis (1982).
[a]Blissymbolics is a system applied to disabled nonverbal children (McNaughton & Kates, 1980).

**TABLE 10.2**
Chronology of Mobility with Reference to Ambulation in Normal and Motorically Handicapped Children

| Age | Walking child | Nonwalking child |
| --- | --- | --- |
| 8–12 months | Crawling, pulling to stand | Castor cart, standing brace |
| 12–24 months | Toddling, sitting and standing control, hands free during walking to carry toys | Castor cart (manual or electric), parapodium |
| 2–5 years | Home, yard and neighborhood mobility | Manual or electric wheelchair |
| 5–10 years | School and community mobility | Manual or electric wheelchair, van or bus for school and community access |

*Source:* Adapted from Motloch (1974).

veloping capabilities. Simple devices can be constructed by family or community volunteers. Robinault (1973), for example, has provided pictures and diagrams for a number of such devices.

**Replacement Assistance**

These devices replace lost capabilities in adults or in those with early-onset disabilities who remain impaired. These are the devices usually mentioned when technology for the handicapped is discussed: wheelchairs, prostheses and orthoses, crutches and canes, environmental control systems, and architectural adaptations such as ramps (Enders, 1984).

**Support Devices**

These devices provide structural support either for comfort or to prevent progression of deformity. In addition, this equipment may facilitate the physical management of disabled persons by a caretaker through use of a hoist, sling, portable commode, and so on. "A study inquiring into physical problems of the caretakers of physically handicapped persons found that 48 of the 49 families questioned reported 'back problems' attributed to lifting or helping the disabled child" (Kohn et al., 1980, 25).

It should be remembered that noninstitutional settings rarely rely on more than one person at a time to assist a disabled person. Therefore, support-assistive devices enable a handicapped person to remain at home for a longer period of time before requiring institutional care. Combinations of assistive equipment, architectural changes, and attendant care may enable a disabled person to remain permanently in a community setting.

## PSYCHOSOCIAL ASPECTS

In order to discuss psychosocial aspects of assistive devices, it is necessary to present a few definitions and concepts about disabilities and technological devices.

### Definitions

An *impairment* is an observable, measurable alteration of anatomy or physiology.

A *disability* is the functional result of the impairment, usually described by the level of performance of daily activities, including work tasks.

A *handicap* is what results from the interaction of the disabled person and the environment and describes the way the individual is at a disadvantage in relation to nondisabled people. Handicaps may be due to external environmental factors that are physical or emotional, including attitudes which underestimate or deny the capabilities or value of disabled people (Wood, 1980).

*Technology* "in its broadest sense is the application of an organized body of knowledge to practical purposes. Rehabilitation engineering is the use of science and technology combined with clinical medicine to improve the quality of life of disabled persons" (Giannini, 1982). Biomedical technology usually refers to devices used in the diagnosis, treatment, or life support of individuals with acute or chronic illnesses. Rehabilitation technology refers to devices providing support, comfort, and/or improved function for individuals with either stable or slowly progressive conditions.

An example of an impairment that does not result in a handicapping condition is seen in a child with amputation of one foot. The loss of the foot constitutes an impairment, but the disability can be removed through the use of an assistive device, a prosthesis, which will prevent the child from being handicapped in his or her interaction with the environment and in usual activities with the peer group. Another example of diminishing the handicapping effect of a disability is the removal of architectural barriers for wheelchair users, thereby allowing disabled people equal access to a building. The impairment has not been cured nor the disability removed, but the handicapping effects in relation to a specific part of the environment have been removed.

The foregoing definitions and examples indicate how technology helps disabled people. One of the problems faced by rehabilitation engineers is the overexpectations of disabled persons and their families for equipment performance and the hope that the underlying impairment will be "magically" cured by the device. When this does not occur, the disabled person and the family may focus their disappointment on the device and on those who constructed or recommended it, and they may reject both the device and the provider. Careful consultation is necessary to reduce this possibility. Families must have a clear understanding of the properties of the equipment, including what it does and does not offer to the disabled person.

**Concepts of Stress and Control**

Stress is defined as "the nonspecific response of the body to any demand" (Selye, 1976). These demands may be pleasant or unpleasant, useful or harmful, and, irrespective of any superimposed specific effects, they may produce a stress reaction. Mandler (1983) indicates that stress situations draw the attention of the individual to these demands and evoke coping responses aimed at mastery or control of the situation. He further states that "it is generally seen as 'good' to be in control, and as the world is appraised as 'good,' the emotional tone will be positive" (201). Thus, it can be said that a sense of control or mastery colors "cognitive interpretation" toward a more positive relationship with the disabled person's environment.

Physical disability creates stress in a previously intact individual. Acquisition of an assistive device may facilitate function, but it may also create stress since it is a new stimulus, which demands an adaptive response toward change. Efforts are then made by the individual to restore homeostasis, a sense of being in balance with one's surroundings, and a self-concept of successful coping.

An example of this effort is reported by Fox (1984), who became a new wheelchair user following the loss of his ambulation as a result of his progressing multiple sclerosis. "The first order of business was to 'make the way' which is how you deal with the turbulence of such a loss. You make the way smooth again, get back to doing things naturally—which meant incorporating a wheelchair into my body image until I could use it as unconsciously as a pencil" (111).

A handicap may be perceived by some as an altered "locus of control" (Rotter, 1966). Disabled persons may feel that control has passed from their own spheres of influence to forces or factors outside themselves. One of the goals of using assistive devices is to return the sense of control to the individual.

An example of this "return of control" was reported by the mother of a 3-year-old, severely disabled boy who was unable to walk or crawl and was therefore immobilized unless someone carried him or pushed him in a chair. When he received an electric castor cart (small wheelchair), he promptly "ran away from home" (around the block) to demonstrate his new ability to control where he chose to be.

Another example is reported by Vash (1983), who was unable to transfer in and out of her wheelchair independently for toileting. She received a sling and hoist, which allowed her to be independent in this activity, and she relates: "The most significant, even startling change was the fact that my dreams changed dramatically within less than two weeks. . . . Although consciously I accepted my dependency (in personal care) with good nature, my dreams showed that it was not a comfortable state of affairs at those deeper levels of consciousness referred to as the 'unconscious'" (48–49).

I do not believe one can overemphasize the importance of providing this sense of control to disabled persons and ensuring their participation in selecting assistive devices.

**Assessment Issues**

Rehabilitation assessment carefully reviews the functional capabilities of the disabled person. Most people want to use their residual capabilities, and these must be correlated with the functions they want the assistive device to accomplish. The environment where the device will be used, the supportive or ancillary equipment that will be needed, the architectural changes that will have to be made, and the type and extent of training that will be required for both the disabled person and others are concerns that must be explored. At the time of assessment, the disabled person and those involved with his or her care should be learning about the available options so they can participate in decision making (Kohn et al, 1983).

Most consumers who have lived with their disabilities for some time have very definite ideas about what they need and what they want the device to do for them. They should be considered the primary partner in the selection and development of equipment, because the better informed they are the more likely the device will be used. However, even with the best efforts of engineers, the device will be a compromise between an ideal system and that which is currently available (Vash, 1983).

Consumer preferences for the color of assistive devices can now be satisfied, since there is a range of available colors in wheelchairs and in coverings for seating and positioning devices. Limb-deficient children and adults can be offered a variety of terminal devices for upper-extremity prostheses, as well as special devices for athletics, musical performance, or heavy farm work. At the Rehabilitation Engineering Center in Stanford University's Children's Hospital, children have chosen colored prostheses ranging from orange to red, white, and blue rather than traditional flesh-colored models. Consumers also "customize" their assistive devices after delivery, using signs, banners, tote bags, and other materials to personalize their equipment, just as people put bumper stickers on automobiles. Providers of assistive devices tend to encourage these activities; they also teach consumers to do simple adjustments and repairs since this helps to increase the person's sense of control and mastery and to incorporate the device into the individual's self-image.

## SUMMARY

Assistive devices are functional-adjunct or structural-support systems that must be incorporated into a disabled person's self-image to be truly successful. They must, therefore, be both cosmetically acceptable and perceived as useful. Effort-reward issues—i.e., is the effort involved in using the equipment repaid by ease of function or additional tasks performed?—can only be resolved by the user of the device. Being in control of the device and, through it, controlling or interacting more successfully with the environment gives both children and adults an increased sense of power. This feeling may reinforce the drive for independence and self-sufficiency.

The stress associated with the new demands of the assistive device presents a challenge to the individual's coping system and will lead to rejection or acceptance of the device according to one's perception of the value to be derived and the resolution of conflicts regarding such issues as dependency. Stress may also be generated by anxiety over equipment failure, repair time, replacement cost, and inadequate sources of financial support. Problems of maintenance, repair, and replacement and the presence of architectural barriers are irritating and may be overwhelming enough to shift the perception of stress toward a rejection of the device. There are deficiencies in both the service system and in reimbursement mechanisms that may severely limit access to appropriate assistive equipment.

There is probably no perfect device, but if the match of the person to the device is a good one, the combination will permit a life-style that is more functional, comfortable, and successful than was achieved without the device. The constant input of knowledgeable consumers is essential to the development of appropriate equipment, and the consumers' and providers' understanding of psychosocial factors is critical to their successful cooperation.

## NOTE

1. The two most important acts of legislation relating to disabled people are: PL 94-142, the Education for All Handicapped Children Act of 1975, and PL 93-112, the Rehabilitation Act of 1973, Title V, Section 504, which states: "No otherwise qualified handicapped individual in the United States, as defined in section 7(6), shall, solely by reason of his handicap, be excluded from the participation in, be denied the benefits of, or be subjected to discrimination under any program or activity receiving Federal financial assistance." (Bill passed Sept. 1973; regulation issued Apr. 1976).

There are two important recent pieces of legislation that will have substantial impact on the availability of assistive technology for disabled people. First is the Rehabilitation Amendments of 1986, PL 99-506 (signed Oct. 21, 1986), whose purpose is "to extend and improve the Rehabilitation Act of 1973." One of the major provisions of the new law requires the inclusion of rehabilitation engineering in the assessment and provision of rehabilitation services. The second piece of legislation is the Technology-Related Assistance for Individuals with Disabilities Act, PL 100-407 (signed Oct. 19, 1988), which is the first piece of federal legislation for the purpose of expanding the availability of assistive technology services and devices to individuals of all ages with disabilities. The act encourages states to establish comprehensive statewide programs of technology-related services, for which it has appropriated $5.2 million for implementation for fiscal year 1988-89.

# REFERENCES

Architectural and Transportation Barriers Compliance Board. (1977). *Resource guide to literature on barrier-free environments, with selected annotations.* Washington, D.C.: Architectural and Transportation Barriers Compliance Board.

Clearfield, D. (1976). *Medical devices and equipment for the disabled: An examination.* Washington, D.C.: Disability Rights Center.

Corcoran, P.J. (1982). Rehabilitation: definitions and concepts. In V.W. Stern and M.R. Redden (Eds.), *Technology for independent living. Proceedings of the 1980 workshop on science and technology for the handicapped.* Project on the Handicapped in Science. Washington, D.C.: American Association for the Advancement of Science.

Enders, A. (1984). *Technology for independent living sourcebook.* Bethesda, Md.: Rehabilitation Engineering Society of North America.

Fox, C. (1984). If you can't stand up, stand out. *Car and Driver,* Oct., 111–121.

Giannini, M.J. (1982). Getting technology to the consumer. Keynote Address, Fourth Annual Rehabilitation Engineering Services Conference, Children's Hospital at Stanford University.

Kohn, J., Enders, A., Motloch, W. & Preston, J., Jr. (1980). *Team assessment of device effectivess* (Monograph). Stanford, Calif.: Rehabilitation Engineering Center, Children's Hospital, Stanford University.

Kohn, J., Enders, A., Preston, J., Jr. & Motloch, W. (1983). Provision of assistive equipment for handicapped persons. *Archives of Physical Medicine and Rehabilitation, 64,* 378–381.

Lifchez, R. & Winslow, B. (1979). *Design for independent living, the environment and physically disabled people.* New York: Whitney Library of Design.

McNaughton, S. & Kates, S. (1980). The application of Blissymbolics. In R.L. Schiefelbusch (Ed.), *Nonspeech language and communication: Analysis and intervention.* Baltimore: University Park Press, 303–321.

Mandler, G. (1983). Emotion and stress: A view from cognitive psychology. In L. Temoshock, C. Van Dyke & L.S. Zegans (Eds.), *Emotions in health and illness: Theoretical and research foundations.* New York: Grune & Stratton.

Motloch, W.M. (1974). Wardrobe of devices. *Inter-Clinic Information Bulletin, 12,* 8.

Musselwhite, C. R. & St. Louis, K. W. (1982). *Communication programming for the severely handicapped.* Houston: College-Hill Press.

Office of Technology Assessment. (1982). *Technology and handicapped people.* Washington, D.C.: Government Printing Office.

Public Law 93-112. The Rehabilitation Act of 1973. 93rd Congress, H.R. 8070.

Public Law 94-142. Education for All Handicapped Children Act of 1975. 94th Congress.

Robinault, I.P. (1973). *Functional aid for the multiply handicapped.* Hagerstown, Md.: Harper & Row.

Rotter, J.B. (1966). Generalized

expectancies for internal vs. external control of reinforcement. *Psychological Monographs, 80* (1, Whole No. 609)

Selye, H. (1976). *The stress of life.* New York: McGraw-Hill.

Sutcliffe, B.C. (1972). Training and education of the physically handicapped: definition of need. *Physiotherapy, 58,* 296.

Tobias, J. (1983). Action, interaction and the thrill of power: Some devices for congenitally disabled children. *Proceedings of the 6th Annual Conference on Rehabilitation Engineering.* San Diego: Rehabilitation Society of North America.

Vash, C.L. (1983). Psychological aspects of rehabilitation engineering. In M.R. Redden & V.W. Stern (Eds.), *Technology for independent living, issues in technology for daily living, education and employment: Proceedings of the 1981 Workshop on Science and Technology for the Handicapped* (Project on the Handicapped in Science). Washington, D,.C.: American Association for the Advancement of Science.

*White House Conference on Handicapped Individuals.* Vol. 3, *Implementation Plan.* (1978). Stock no. 040-000-00395-9 Washington, D.C.: Government Printing Office, June.

Wood, P.H.N. (1980). Appreciating the consequence of disease: The international classification of impairments, disabilities and handicaps. *WHO Chronicle, 34,* 376–380.

# 11

# Sexual Concerns of the Physically Disabled

*Susan E. Knight*

The sexual concerns of the disabled began to be addressed in the early 1970s at the same time that human sexuality became a major topic of interest in the helping and lay communities. Initial interest focused on spinal cord–injured persons largely because of the significant number of Vietnam era veterans who returned from the war with this disability. As these men began to demand their human and sexual rights, consumer and advocate groups interested in other disabilities also began to speak of their need for sex education and counseling. During the 1970s, several major medical centers in the United States and abroad began providing counseling and training in sex and disability. Clinical work conducted in these settings and in the community resource centers for the disabled led to an understanding of the major sexual concerns of persons with disabilities. This chapter will delineate the clinical problems and the interventions that have been successfully used to treat them.

Understanding the sexual concerns of persons with physical disabilities is a difficult task. It is crucial to remember that disabled persons and their partners are men and women first and disabled second. The sexual effect of the disability varies dramatically from person to person and from couple to couple. Because of this diversity of personal experience, it has been difficult if not impossible to develop step-by-step interventions to remedy the sexual problems of this highly diverse population. However, there are several specific concerns frequently reported by the disabled. Most widely reported are concerns about sexual discrimination and body image; from the outset these must be dealt with in an empathetic and focused manner.

Persons with disabilities form a heterogeneous group because each person's age, prior life, and sexual experience greatly affect his or her adjustment to disability. A person born with a disability will usually have a very different sexual awakening from a person disabled in adulthood. The individual's personal and sexual development is more clearly affected by the exact degree and type of disability and the availability of family and peer support, as well as the person's age at the onset of the disability.

Twenty years ago, a person with a disability lived in a society that basically saw him or her as radically different from others, unattractive, and as someone who did not experience sexual feelings. In early work at the Sex and Disability Unit of the Human Sexuality Program at the University of California, San Francisco, it was not uncommon to counsel persons with disabilities who wanted to know why they *were* feeling sexual. Not only were stereotypes of asexuality held by the society at large, but often by the disabled individual as well. The civil and social rights movement of the disabled in the 1970s has done much to erode this and other myths about disability, and today most persons with disabilities know that they are sexual human beings.

The universals of sexual feeling are increasingly acknowledged by health and human sciences personnel and even to some, though lesser, degree by society, because of such mass entertainment vehicles as the movie "Coming Home." However, the recognition of one's sexuality by oneself and others is only the first step in the process of developing social and sexual skills, selecting a partner, and dealing with continuing sexual discrimination.

Of great importance are the feelings of the actual or potential partner, whether disabled or able-bodied. If I have a disability, how do I feel about other disabled persons as potential partners? If I am able-bodied, what attitudes about disability do I bring to this relationship? If the person I am in a relationship with has significant mobility restrictions, what role do I play in that person's life? Am I lover, personal care attendant, or both? What do I feel about possible restrictions on our social and sexual lives? If I am either disabled or able-bodied, how do I respond when my family and friends question my choosing a disabled person as a partner? These issues are central to the therapeutic process when dealing with the individual and couple and need to be addressed together with concerns of sexual enjoyment and pleasure.

## CONGENITAL DISABILITY

For most able-bodied parents and siblings, the birth of a child with a disability is often their first personal exposure to disability and to what it feels like to be seen as different. Parents or siblings who show the child to family and friends or bring the child into public places are often confronted by reactions of pity, repulsion, inappropriate curiosity, sadness, and even anger. Since the child is an extension of themselves, family members experience firsthand the isolation and at times the shame that come from being "not normal" or handicapped.

When a child with a disability is born, the parents experience the loss of the "perfect" child they had hoped for. Because the child with a disability is not dead, however, but alive and disabled, the child is a daily reminder of the loss they have suffered (Solnit & Stark, 1961). If the parents see their feelings as a healthy response to having a child with a disability, they will be able to accept these feelings and go on with the tasks of life. Parents who experience their lives and that of their child as mainly tragic will not experience their child's sexuality—or their own—as a positive and natural part of life. Certainly one way to insure positive sexual growth of congenitally disabled children is to encourage the sexual rights of their parents. Helping parents find time for themselves and their relationship and providing sex education and counseling for them will go far in preparing them to see their children as sexual.

Establishing mature sexual relationships demands that one achieve a significant degree of independence from parental figures. The person born with a disability often remains physically dependent on parents much longer than the able-bodied individual. The parents of the adolescent or the young adult with a disability must not only condone the independence of their child but, because of mobility problems, must encourage the child's social life and actively participate in facilitating it. The parents of the able-bodied child know that the child will most likely "take" independence. Parents of the child with a disability such as cerebral palsy must actually give and encourage the child's independence. These parents are often in a dilemma when it comes to the social norm that says a person can be sexual only when he or she becomes independent of parental figures. Given this societal value, it is not surprising that many parents who want the best for their child often have great difficulty helping the child obtain the social and sexual information and experience it needs. Because of these societal expectations, it is common for people with congenital disabilities to seek sexual counseling after they become physically independent of their parents.

A 40-year-old man with moderate cerebral palsy came to see me for sexual counseling. When asked why he came for counseling, he said, "I think I'm old enough to learn about sex." He was college educated, fully employed, and had been living by himself, away from his parents, for 3 years. In questioning him further, I found that the extent of his sexual knowledge was that he knew he had a penis but knew nothing about the human sexual response or about female anatomy. When I asked him if any "white, sticky stuff" ever came out of his penis, he replied, "Yes, and doesn't that have something to do with my cerebral palsy?"

This man had obviously got the message early in life that sexuality was something only able-bodied people experienced and that since he wasn't "normal," all the sexual messages and information that surround everyone didn't apply to him. Two things in this man's experience stand out as common to the sexual awareness of congenitally disabled persons:

**1.** Independence from parental figures usually allows individuals to ask questions about sexuality for the first time, regardless of age.

**2.** Prior to adequate sex education, individuals often link sexual feelings and experiences with the disability rather than with pleasures that are shared by all people. At this stage of awareness, people often view themselves as "a disability" rather than as persons with disabilities.

Therapy for this man consisted of having him purchase a book on sexuality and discuss the information with me. He became excited to learn that his feelings and sensations were normal and soon acknowledged that he masturbated to orgasm, an experience he had had no label for previously. Once he had information about female anatomy and sexual expression, he became very interested in what a sexual experience with a woman would be like, as well as how he might meet women as either friends or lovers. He was socially isolated and required counseling to develop social skills and to learn of possible ways he might gradually increase contact with women, as through coffee dates, church and social functions, and classes where he might meet interesting people. He was also encouraged to begin attending meetings and social events within the disabled community. Not only did this man not know that his sexual responses were normal, he also did not realize that his sense of being different was shared by many of his disabled peers. The development of a sense of group identity with other disabled persons is often a necessary step in realizing that one is human first and disabled second.

This man's next step was to gain some sexual experience so that he could become familiar with what sex with a partner felt like as well as what part his disability played, if any, in his sexual expression. A significant number of virgins with disability have difficulty in the early development of sexual relationships because they fear they will not know how to perform sexually or wonder if they can function sexually at all. By gaining initial sexual experience in a supportive environment, such as sessions with a sexual surrogate, questions like "How can I?" can be answered, freeing the person to initiate contact with potential sexual partners. A sexual surrogate is a counselor or trained paraprofessional who does sexual homework assignments with clients who do not have an available partner. The work is performed under the guidance and prescription of the sex therapist. This man was seen by a surrogate for five 2-hour sessions. During these meetings, he was able to discover the sexual experiences he enjoyed with a partner and learn how female genitals look and feel as well as how to touch and kiss a partner and to ask how she would like to be pleasured. Supportive, therapeutic sexual experience with surrogate partners does not entirely prevent fears of performance in future, nonclinical sexual encounters; however, this kind of bodywork therapy can help answer the question "How can I do it?" and can thus be critical to the sexual development of persons with disabilities, who have limited access to initial sexual experience.

Some therapists, especially those outside large metropolitan areas, may not have access to the services of trained sexual surrogates or may wonder if this therapeutic learning might not be obtained through experience with prostitutes. The answer is, most often, no. One man with mild cerebral palsy came for counseling because he could not develop close social relationships with women. He had been sexually active with prostitutes for 5 years. He felt that he could not develop a relationship because he did not make enough money (he was fully employed and lower middle class) and because he felt he ejaculated too quickly for a woman to accept him sexually. The feedback he received from a number of women, including the therapist, was that he aggressively approached them as sexual objects. He soon left therapy, admitting that it was too anxiety provoking for him to talk about his feelings, especially in the presence of a woman he considered attractive. He said he used to deal with this anxiety by "getting off and getting out." His experience with prostitution—of paying for sex with the expectation that he was there to ejaculate rather than to relate interpersonally—had given him the message that unless he was wealthy and could last a long time he deserved no more than a "quickie" and could not be attractive to a "real" (nonprostitute) woman.

This client's experience points out not only the limitations but, more importantly, the potential destructiveness of exclusive and sustained sexual experience with prostitutes for most men. A business where sexual response is timed and charged and feelings are ignored and repressed is not a place where sexual or personal growth can occur. It is not surprising that this man equated paying for therapy with prostitution, so that paying for therapy meant, to him, that the therapist could not possibly have the client's best interests in mind. With his image of how a man acts toward a woman he pays, he continually approached the therapist sexually and expressed in many ways that the only way she could prove that she saw him as a person rather than a customer was to drop the therapeutic relationship and date him socially and sexually. This man's experience is truly tragic; he had been conditioned, partly because of his own naivete and low self-esteem, to be continually frustrated and sexually limited in his relationships with women. A society that says, "Well, at least he can get off with a prostitute," is damning this person to a restricted and unfulfilled sexual and social life. "You get what you pay for" does not easily translate into the currency of intimacy and sexual growth.

It is not surprising that this discussion about gaining sexual experience has focused on men. Culturally, we have expected men to gain sexual experience before developing an intimate relationship. While "sowing wild oats" is less of an expectation for men today than it once was, it still exists in the social and sexual repertory of American men. Likewise, the idea that a woman, able-bodied or disabled, would seek out a therapeutic relationship for the purpose of exploring her sexual response with a partner is seen by most people as very strange indeed. Male and female surrogates who work with women do exist, and their services are equally valuable; however, they are used much less frequently than their counterparts for men. Most women with disabilities rely for their initial experience on either the development of a committed relationship or, more commonly, the approach by a willing but casual partner. While casual can mean caring, pleasing, and even educational, more often it is experienced as a "sympathy fuck"—"She is only doing this because she doesn't think anyone else will." Women who experience the feeling that "I'd better take what he is offering, even if I don't particularly like him" experience a self-devaluation not too dissimilar to that of men with prostitutes.

Society stereotypes disabled persons as passive. Because of this expectation, many people with disabilities are targets of incest, sexual molestation, and rape. While current studies are incomplete, it is believed that disabled children and adults are at greater risk from sexual abuse than are the able-bodied. Thus, the role of the disabled as social victims becomes actualized in the act of sexual abuse. When viewed in retrospect, the early sexual experiences of many disabled persons are described as acquaintance rape. Given the societal expectation that disabled persons should not resist the advances of persons who are sexually interested in them, it is not surprising that many disabled people feel sexually exploited by not only the individual involved but by society as well. This experience of victimization is shared by both men and women, gay and straight.

Persons with disabilities, especially those disabled from birth who may never have experienced making choices in relationships, need counseling in assertiveness, making choices, and sexual exploration through masturbation. Such experiences allow individuals to learn about their own sexual responses and to see sex with a partner as an option at any point in their lives rather than the only way to be sexual.

## ACQUIRED AND TRAUMATIC DISABILITY

The person who becomes disabled after spending childhood and adolescence as able-bodied grieves over the loss of a body part, a degree of mobility, certain life experiences, and certain relationships with friends or family members for whom the newly disabled individual is "not the person I used to know." Acquired disability covers a vast range of physical losses and experiences, from surgery and traumatic injury and disease to the aging process. The loss associated with an acquired disability is as great as the person experiences it to be and is not necessarily tied to the specific degree of physical impairment or disfigurement. Not surprisingly, persons who have had positive relationships with disabled people before their own disabling experiences often put the disability in a more positive perspective than that of people who have not been exposed to this area of life in the past. Sexually, the same holds true.

Persons who experienced sexual difficulties or great sexual anxiety before their injuries will most likely see a disability (whether or not it actually affects sexual mobility or response) as a major hurdle or the "death blow" to their sexual experience. Similarly, individuals with a satisfactory but limited range of sexual expression prior to the loss may now feel themselves sexually inadequate. (For example, a male leg amputee who had only had intercourse in the "missionary" position now finds that he is better balanced in the side-to-side or woman-on-top position.) Sex therapy for the person and couple experiencing acquired disability needs to explore the impact of the new disability on each person. This examination should address what has and has not been changed physically and emotionally and how these changes relate to the couple's sexual activity. In the case of a change in physical mobility and/or sexual response (e.g., in spinal cord injury), increasing the number of sexual activities is often necessary for satisfactory sexual expression.

In a relationship where one or both people are spinal cord injured, an individual with a disability can often have coitus as well as stimulate a partner through oral and manual means. However, a noninjured partner may feel frustrated not being able to sexually satisfy the injured partner. The impact of spinal cord injury on the sexual response cycle is described elsewhere (Geiger, 1981). Often the more important aspect of how a person with limited sensation can gain sexual satisfaction is much less explored in the literature. Through self- and partner-sensate focus exercises, a number of persons with spinal cord injuries have found areas of their bodies (the breasts, the area where sensation changes from full sensation to decreased sensation, and the ears and neck are common sites) that, when stimulated, lead to successful completion of the sexual response cycle. The Masters & Johnson Institute provides an explicit example of this physiologic phenomenon in a paraplegic woman without genital sensation, during masturbation in the laboratory:

When the subject was monitored . . . during breast stimulation, it was noted that cardiopulmonary responses to sexual excitation and orgasm were normal and changes observed in the breasts and nipples were normal, but there was not significant degree of pelvic vasocongestion or vaginal lubrication. However, in the late portion of the plateau phase of the response cycle (identified by breast changes that occurred), the lips of the woman's mouth became engorged to twice their normal size. At the moment of orgasm, a pulsating wave was observed in her lips and the swelling then dissipated rapidly in a manner almost identical to the pattern seen with the dissipation of the orgasmic platform formed at the outer portion of the vagina in non-cord-injured women. (Kolodny et al., 1979, 368)

To date, no systematic treatment plan of sensate exercises has been published for persons with limited or no genital sensation. However, the model of group or individual treatment of nonorgasmic women using self- and partner-sensate focus exercises described by Barbach (1980) gives useful guidelines for developing treatment exercises for spinal cord–injured men and women. While these individuals are not anorgasmic in the classical sense, the need to develop sexual response in a previously understimulated area can be seen as analogous to women who have not experienced full genital arousal and response.

A note of caution is warranted here for the clinician unfamiliar with the range of disabilities that cause limitations of movement. It is assumed by many that anyone who uses a wheelchair must have limited genital sensation. While a number of persons with spinal cord injury, spina bifida, and in some cases multiple sclerosis do experience changes in genital responsiveness, the majority of disabled persons who use wheelchairs have complete genital sensation and unchanged genital sexual response. Even among persons with changes in genital sensation, many have some degree of genital sensitivity and awareness remaining, which they may or may not know of themselves and which may possibly be incorporated into their sexual repertoire. Sensation and response can be a separate phenomenon in this population. Up to 85% of spinal cord–injured women lubricate vaginally when stimulated directly, and the same percentage of men get erections whether the individual is aware of the sensation or not (Geiger, 1981). Therapy should include assignments to the individual and couple to explore these responses and find out what degree of sensation and responses does exist and how it can be incorporated into the sexual repertoire.

## ISSUES OF PARTNERS

The partner of the person with a disability has often been neglected in therapeutic work. When a person acquires a disability, the partner will often face many of the same feelings and social reactions that the parent of a child born with a disability does. Partners, who most likely have no experience with disability, are now flooded by negative stereotypes about disabled persons, including the feeling that their loved one must now be asexual, dependent, and emotionally inadequate. At the same time, if partners are successfully able to label these messages as untrue, they will still meet with pity or morbid curiosity from family, friends, and strangers, who speak about how noble, crazy, or unaware they must be to remain with the disabled person after the injury. They may also be receiving these messages from their partners, who especially in the early stages of dealing with the disability may feel they can no longer fill their social, intimate, and sexual roles. While the sexual relationship issues of the partner are often disregarded by social workers and other helping professionals, they do need to be dealt with, in both an individual- and couple-counseling format (Hartman et al., 1983).

Whether or not the couple was in the relationship when the disability began, the issue of what physical roles the able-bodied (or less disabled) person will assume can greatly affect the sexual relationship of the couple. If one partner has to do all the physical tasks around the house, the other should take on activities such as managing financial or social/recreational activities so that both partners feel an equality of effort. Of course, hiring others to take on some of the household duties can be an attractive alternative for partners with financial resources. Not surprisingly, these issues often need to be negotiated in couple counseling. These are issues similar to those brought up by able-bodied couples under the heading "Who's going to clean the bathroom?" An astute counselor will notice when one or both partners are negative about hiring someone else to help because of a feeling that the less disabled partner should do all physical tasks "if he or she really loves the other person." This attitude has to be challenged; inequality does not help sustain love. Many able-bodied couples would be delighted to hire someone else to do unwelcome chores. An associated issue is how much personal care (dressing and toileting) one partner should give to another with severe mobility impairment. Some people feel that being both an attendant and a lover has a negative effect on the sexual relationship and find that one or both people have difficulty switching from a clinical perspective to a more intimate, physical relationship. Still others find this kind of interaction and even the physical ramifications of the disability to enhance their sexual expression, as in the following vignette written by a spinal cord–injured woman, whose lover serves as her attendant.

A sense of humor is so very necessary in our relationship; we goof about everything that goes on. You can't get too serious about incontinence. We allude to it in our lovemaking; we talk about "golden showers." Even though we empty my bladder before intercourse, there's no guarantee that I won't be incontinent while we're having intercourse because of the stimulation of pushing on the bladder. If I should wet, well, that's a normal and exciting part of our relationship. (Finkle et al., 1981)

Couples need to be reassured that their preference for one style of attendant care is not an issue of love but of preference. They need to remember that barbershops and laundries exist for people who choose not to do each other's hair or shirts and that the same goes for the preference of hiring outside attendant care when available and affordable. Other issues, such as fixed limits on shared physical activities (also an issue of able-bodied couples, usually labeled as individual preference rather than physical necessity) and the ability to hire outside help, should also be explored. Such inquiry is important to see whether further steps would increase equality and decrease real or potential frustration and unresolved anger in the relationship.

The theories and techniques of current sex and communication therapy are vital tools for working with individuals and couples affected by physical disability. Persons with physical disabilities and their partners experience sexual and communication difficulties not related to the disability. Such concerns, and those that are specifically related to the disability, can be handled in a supportive, educative, behavioral, and short-term fashion, just as they are with able-bodied individuals and couples.

Other issues such as sexual preference, sexual attitudes, AIDS prevention through safe sex, sexual and physical abuse, and alcohol and drug abuse can have a great impact one's sexual experience and adjustment. This is as true for the individual with a disability as it is for the able-bodied. These factors are handled in the counseling process along with the topics specific to persons with disabilities.

## SOCIAL SKILLS

Social skill problems are common for the individual with a disability who desires but has difficulty obtaining social or sexual relationships. Fear of rejection, questions of how to present the disability to a person one is interested in getting to know, and what to look for in a potential partner are issues that are shared by single persons with disabilities.

Many persons with disabilities anticipate more rejection than they actually face. The person born with a disability may have experienced rejection by able-bodied peers during the school years, which led to the expectation, an overanticipation, of continued rejection. The person who acquires a disability may paint a mental picture of an isolated and bleak social future. Even as therapists need to acknowledge to clients the existence of sexual and social discrimination toward disabled people, they must point out that as a result, their marital status or other relationships may not inform a correct measure of their self-worth and mental health. However, disabled individuals' stereotypes about how others view and treat them should be explored to make sure that they are not acting inappropriately, thereby fostering rejection for reasons unrelated to disability. Clients need to be questioned to see if they are withdrawing socially, thus avoiding any chance of either rejection or acceptance. Asking questions such as, "Tell me about the last time you remember being rejected?" will give valuable information on how much the clients interact with others or whether the clients themselves are interacting in a rejecting or overly aggressive manner. Some people seem to mirror the rejection they expect; they reject others through conscious or unconscious means. Another indication of social development can be measured by a person's nonsexual relationships. While it is true that persons with disabilities may not always have potential or real sexual partners, most can have successful and close friendships with disabled and able-bodied peers. Individuals who portray themselves as unlikable need to develop friendships before progressing to greater degrees of intimacy and its attendant vulnerability. Books on social skills development, such as *Shyness* (Zimbardo, 1977), are as applicable to a person with a disability as to the able-bodied. Assigning a shy individual to speak to three people each day or ask someone out for coffee in order to get acquainted may enable appropriate social interaction.

Two common social mistakes of both disabled and shy people are: (1) wanting to spend time with anyone of the appropriate sex and age range who will spend time with them ("I don't have any personal preferences and certainly wouldn't reject anyone's interest in me") and (2) asking people out for dates that are too long. Most people will spend 15 minutes at coffee with friendly acquaintances but, unless very attracted to them, will turn down an invitation for the entire evening or day. Social skills exercises need to maintain the comfort of both persons, or the person with the disability will invariably face rejection because of his or her style of relating to people. The individual needs to learn enough patience to allow a relationship to develop slowly enough so that each positive interaction builds on the previous one.

Another reason that disabled people experience rejection is because of the universal discomfort with disability itself. People who are socially skillful learn how to speak about their disabilities in a way that decreases others' anxieties and shows a sense of positive self-esteem. One client with a prominent speech problem from cerebral palsy was assigned to answer others' questions about his speech, not by describing cerebral palsy or by talking about its negative impact on him but rather by stating, "I have a speech disability that affects how I pronounce certain words." This exercise allowed him to contain the social effects of his disability in this situation. His statement implied that while he had a difficulty, it was only with certain words. This man certainly needed to speak further about his disability when he made longer and more intimate social contact; however, he had begun to learn to fit the explanation of the disability to the particular situation rather than the other way around.

Such containment of disability also allows people to present themselves first rather than their physical limitations. When one woman was approached by strangers who told her they were sorry she had a disability, she said, "I looked them in the eye and, smiling in a friendly way, said, 'Thank you, but it has added more to my life than it's been a problem.' I usually excused myself at that time saying, 'Sorry, I can't talk now, I have to get back to work.'" Such polite but purposely brief responses indicate a solid sense of self and also reflect a personal sense of priority (i.e., "I'm a busy, productive person, not the 'victim' you might imagine me to be"). These positive, nonstereotypical responses serve two purposes: (1) they challenge stereotyping and provide a pleasant surprise when the person with the disability does not react in an expected way and (2) they increase self-esteem by inducing one to talk positively and nondefensively about oneself. Clients need to practice answers to public questions, both to insulate themselves from others' negative attitudes and to increase their own sense of well-being.

Disabled people often ask, "Given the social expectation that disabled people are asexual, how do I let people know I am a sexual person?" The way not to do this is to sit with a person one is sexually interested in and mention the word "sex" for the first time by saying, "I'm really nervous about saying this, because you probably won't like this, but I want to have sex with you." While being painfully honest, this approach is asking for rejection. To interact successfully, sexually or otherwise, we must project those qualities we want to share with others. To let others know that we are sexual, we need to communicate that to people in general, not just to those few people we are specifically interested in. By dressing and carrying ourselves in a sensual way (as best as our disability will allow) and by appropriately discussing our sensual and sexual preferences (e.g., describing preferred types of touching and the people we find particularly attractive), we let others know that we see and express ourselves sexually. Increasing the amount of personal and sexual secrets we share with a person we are attracted to sets the stage for possible sexual interaction.

The question "Will you be sexual with me?" still often needs to be asked. However, if it is brought up following increased intimate communication, there is a greater possibility of the other person accepting the proposition. As part of the agreement to be sexual together, the individual with a disability should share any information not previously voiced about the sexual ramifications of the disability. Also, partners should be asked if they have any questions or concerns and should be questioned about their own sexual responses and preferred activities. While this amount of communication seems cumbersome and certainly not spontaneous, persons with disabilities do not have the luxury to be "swept away" by passion without communication. While seeming to be mechanical, this kind of honest communication of preferences and physical limitations sets the groundwork for good sexual communication on a continuing basis.

Unfortunately, most persons with disabilities encounter potential partners who care deeply about them but can't get past their negative feelings about disability to be involved sexually. This can happen during the couple's first sexual encounter when the partner enters the sexual relationship hoping to overcome uncomfortable feelings about disability. Therapists can encourage clients to act out such possibilities (as well as other highly charged sexual situations, such as incontinence during lovemaking) so that a disappointment does not have to be something that "I never even though could happen." The sense of helplessness—and often anger—that persons with disabilities experience and that is associated with not being able to change the negative attitudes of people they care about, is very common and should be accepted by the therapist with empathy. It is important not to use phrases such as, "None of us are acceptable to everyone," or, "It's like racial discrimination." Sexual discrimination is not racial discrimination. People with disabilities (except deaf people who use sign language) do not have a subculture in which they are openly accepted. Even among people who are disabled, there is much discrimination against individuals and subgroups. The client's sadness needs to be expressed and accepted. Once the pain of the specific event has somewhat dissipated, the therapist should discourage the client from trying either to convince the desired person to have sex or to spend more time together than is appropriate. While one certainly does not need to throw away a good friendship if a loved one does not want sex, spending a great deal of social time with that person, and thus not meeting other potential partners, is self-defeating. Many people in these situations settle for the attention of the loved one at the cost of their own desires and wants. This compromise should be continually challenged by the therapist, and the client should explore plans to meet other potential partners.

All clients need reminding that their particular social and sexual concerns are concerns shared by many other people, both able-bodied and disabled. Persons with disabilities need encouragement to develop their skills in the mainstream of society rather than in special programs for the disabled, unless the individual so desires. For example, persons who have concerns about body image (e.g., "Will people feel I'm repulsive?") and their abilities to give sensual pleasure can be encouraged to attend workshops or classes on massage with either able-bodied people or a combination of able-bodied and disabled individuals. In anticipating involvement in such classes, many people will have concerns about their acceptance by others, the physical accessibility of bathroom facilities, and so on. The therapist will need to put on his or her "social-work hat" to help the client find a way to gain access to these activities and experiences.

In speaking of dating skills, it is important to ask, who are the partners of the disabled? Clinical and personal evidence points to the theory that potential partners are often individuals who have had a positive relationship with a disabled person, usually during childhood. Therefore, siblings, children, nieces, nephews, cousins, and the next-door-neighbors of persons with disabilities are all potential partners. The exact nature of the person's disability seems not to be important. The fact that this was an important person in the potential partner's life, who happened to have a disability, seems most significant. Individuals who have had positive experience with disability seem to easily accept the idea that a desirable person can have a disability. It is appropriate for disabled persons who are seeking relationships to casually inquire into potential partners' experience with disability. While this should certainly not be used to totally exclude any one individual, it can give an indication of a person's initial comfort with the issue of disability. Clients can also be encouraged to stop or limit the development of relationships they do not really desire or in which they feel the other persons' attitudes about disability are counter to their self-esteem. Clients need to know that they do not have to accept another's feelings or attitudes at their own expense.

## ISSUES FOR PROFESSIONALS

Helping professionals interested in providing sex counseling to people with disabilities have some specific work to do before they begin. First, they need to examine their own attitudes about both sex and disability and to gain some degree of comfort in discussing the issues involved. Working with centers that serve the disabled (especially if they are run by consumers) can do much to help one face personal attitudes about the social and sexual lives of disabled people. Second, it is important to learn and use the clients' knowledge of themselves and their particular situations to come up with creative and workable solutions. Each person's disability affects each one differently, and there is no standard of normal disabled behavior. Individuals should be encouraged to solve problems for themselves with the support of the therapist. In the same regard, specific information gathered from reading about sex and disability should not be applied too strenuously. It is important to share such information openly with the client and talk about its applicability to this person's experience.

Lastly, professionals who are not themselves disabled ought to find resource people with disabilities, who can provide peer counseling and information and referral to clients. As mentioned, interaction and consultation with other persons and partners experiencing disability can be very important to the client's sense of self-respect. While counseling from a disabled therapist should not be pushed onto clients, it needs to be brought up as a possibility, especially at times when clients express feelings that they are the only ones having a particular experience. Peer counseling at such times can be invaluable.

## SUMMARY

Although the past 18 years have given us much information on the sexual challenges confronting the disabled, there continues to be a need for studying the sexual concerns of this diverse group and for satisfying the intimacy needs of disabled individuals. While we cannot hope to develop therapeutic techniques that will address the sexual concerns of all persons with disabilities, we must learn to systematically describe interventions that have worked well with individuals in this population. The greatest challenge is to society at large—to allow the person with a disability true access to intimacy and sexual expression.

## REFERENCES

Barbach, L. (1980). Group treatment of anorgasmic women. In S. Leiblum & L. Pervin (Eds.), *Principles and practice of sex therapy*. New York: Guilford, 107–146.

Finkle, P., Fishwick, K., Nessel, K. & Soliz, D. (1981). Sexuality and attendant care. In D. Bullard and S. Knight (Eds.), *Sexuality and physical disability: Personal perspectives*. St. Louis: C.V. Mosby.

Geiger, R. (1981). Neurophysiology of sexual response in spinal cord injury. In D. Bullard & S. Knight (Eds.), *Sexuality and physical disability: Personal perspectives*. St. Louis: C.V. Mosby.

Hartman, C., MacIntosh, B. & Englehart, E. (1983). The neglected and forgotten sexual partner of the physically disabled. *Social Work,* 28(5), 370–374.

Kolodny, R., Masters, W. & Johnson, V. (1979). *Textbook of sexual medicine.* Boston: Little, Brown.

Solnit, A. & Stark, M. (1961). Mourning the birth of a defective child. *The Psychoanalytic Study of the Child, 16,* 533.

Zimbardo, P. F. (1977). *Shyness.* Menlo Park, Calif.: Addison-Wesley.

# 12
# The Forensic Psychiatrist's Role in Evaluating Physical Disability in the Workers' Compensation System

## Carroll M. Brodsky

Forensic psychiatry overlaps two cultural sets, the legal and the psychiatric. Each has its own rules, roles, practices, beliefs, ethics, and values. The medical anthropologist is fascinated by the diffusion of culture and the acculturation that occur in this area of overlap and how these affect lawyers and psychiatrists when they return to their own fields.

This chapter will discuss differences between the clinical psychiatrist, who diagnoses and treats patients and families, and the forensic psychiatrist, who examines *applicants, claimants,* or *plaintiffs* but does not treat these examinees. Instead, based on the examination and related records, he or she renders opinions to lawyers and courts about any existing disabilities and their causes. The chapter will focus on the forensic psychiatrist's approach to cases of disputed disability, the issues raised in such cases, and the problems of dealing with those issues, especially when physical and psychiatric disability overlap. The discussion is oriented to disability evaluation in workers' compensation systems and is based on my experience in evaluating thousands of such cases.

There are many physically disabled persons in our society. The number varies with the definition of *disability,* which might include those whose disability predates birth or is caused by perinatal trauma and those who become disabled as a result of accident, infection, or late-appearing genetic effects after a period of normality. Society recognizes that it must assist these disabled persons, with private charitable groups and legislatively established governmental organizations serving as agents. Private insurance provides monetary and medical support for disabled persons, and this insurance is purchased by individuals or their employers. Because anyone causing harm to another person can be held liable for the damages, individuals and corporations purchase insurance to protect themselves against the costs of the suits and the losses that might result from judgments against them. In each state, the law requires employers to insure themselves so that they can provide benefits for disabilities arising out of or caused by work-incurred injuries and illnesses (Larson, 1978).

Forensic psychiatry plays its role when there are disputes between the person claiming disability and the persons or agencies that might have to compensate them. Such disputes are presented to the following tribunals:

- Social Security administrative law courts that consider entitlement to benefits from Social Security Disability Insurance and Supplemental Security Income
- Federal and state workers' compensation boards that deal with medical and rehabilitation benefits for disability
- Civil courts that deal with tort claims, including personal injury and malpractice
- Other forums that deal with such issues as disability retirement, veterans' disability benefits, and the claims of those who believe they have been wrongfully disqualified from work on the basis of disability

The issues disputed are
- The presence of disability
- The degree of disability, if its presence is conceded
- The causes of disability
- The degree of pain and suffering the disabled person has experienced
- The proper forum in which the dispute should be heard

Disputes are raised
- To protect public funds, such as Supplemental Security Income or Social Security Disability Insurance
- Because private persons or their insurance companies believe they did not cause the disability
- Because employers disagree that the disability is due to the job
- Because those claiming disability have notions of entitlement different from those who would pay them
- Because some lawyers believe that they must test the limits of the system in order to do justice to their clients or to avoid malpractice claims against themselves

- Because we live in a litigious culture in which many people believe they are being cheated by whatever system they are in (Galanter, 1983)
- Because sometimes health care professionals tell patients their disabilities were caused by injuries when, in fact, they were the result of spontaneously occurring and evolving diseases

## THE WORKERS' COMPENSATION SYSTEM

Prior to the enactment of workers' compensation legislation, a worker injured on the job had to prove that the employer had caused the injury willfully or through negligence and that the injury was not due to the negligence of the worker or coworkers. It was a "fault" system. Disputes were resolved in the same courts that heard tort cases concerning willfully and/or negligently inflicted harm (Larson, 1978). In contrast, workers' compensation laws are "no fault." Any worker who suffers an injury or illness that "arises out of" or in the course of employment becomes a beneficiary. These statutes do not require proof that the employer did any wrong or that the worker did not (Larson, 1978).

The workers' compensation system consists of state and federal laws, which provide income replacement, medical expenses, and rehabilitation services to individuals disabled by work-related injury or illness. The programs of the 50 states and those of the federal government vary, and a specific program is comprehensible only within the socioeconomic framework of its locale, the legislative history from which it arose, and the case law interpreting its provisions (Heiman & Shanfield, 1978).

Workers' compensation systems seek to rehabilitate an injured worker and enable him or her to resume a productive role in society. A disability is compensable when it totally or partially limits a worker's ability to compete in the labor market. There must be a decrement in earning capacity, and that decrement is calculated not only in terms of the physical disability itself but also in terms of age, education, and previous skills. In most instances, disabled workers are paid more when they are recovering than when no further significant recovery is likely. Medical conditions arising out of work are treated at the expense of the employer, even in the absence of disability.

Obtaining workers' compensation benefits is usually an administrative procedure. Ninety-two percent of claims filed in California are not litigated. Benefits are paid to the injured workers or to their survivors if the injury resulted in death (Witt, 1985).

Although the workers' compensation system is "no fault," disputes do arise, beginning with the fundamental questions of whether there is an injury, illness, or disability and if so, whether it arose out of the employment. These disputes are heard by referees or judges presiding over administrative courts. There are channels of appeal to higher administrative bodies and even to state supreme courts and the United States Supreme Court (Harter, 1983–1984).

## THE FORENSIC PSYCHIATRIST

Dispute-resolving legal forums often deal with issues beyond a layperson's knowledge. Therefore, they bring in experts to help them understand the technical aspects of the issues and to assist them in making their decisions. In the American dispute-resolving system, contending sides bring in their own experts to help the judge and/or jury understand the facts and to convince them to resolve the conflicts favorably (Bazelton, 1974). Sometimes judges bring in their own consultants.

Psychiatrists who function as experts in dispute-resolving fields are known as forensic psychiatrists. They testify on issues of sanity, competence, impairment, and disability—according to definitions established by legal agencies and accepted for this purpose by the medical profession. In claims alleging injury to the psyche, the forensic psychiatrist assesses the mental state of the applicant and determines if mental impairment and disability exist. Forensic psychiatrists can be retained by one side, as in the adversary system, or they can serve as Agreed Medical Examiners or Independent Medical Examiners, in which cases they are presumed to be impartial.

Judges in the workers' compensation systems must weigh conflicting psychiatric opinions. These may result from a lack of objective external or physical criteria to validate a diagnosis of some mental disorder, or because the patient's condition is evaluated at different stages of the claimed disability, in different contexts, or because mental health professionals have different kinds of training and experience and adhere to different theoretical schools of thought. Further, each professional is subject to personal biases (Gorman, 1983).

The causation of a mental disability is multifactorial and comprises many diverse contributing factors: heredity, past and present social and cultural environment, interpersonal experiences, age, physical health, marital and social status, and occupation, which may in itself reflect many of the other factors. Therefore, determining that a workers' employment has caused mental impairment or disability may be difficult (Griffith, 1983; Heiman & Shanfield, 1978). Psychiatrists are also asked to predict the course of the mental disorder; if the job is judged to be its cause, the benefits paid will vary according to its extent and expected duration and will depend on what treatment will be required.

## THE SPECIAL CULTURE OF FORENSIC PSYCHIATRY IN WORK-CAUSED DISABILITY

### Diagnosing Physical Disorders

In contrast to their approach to mental disorders that do not involve physical disability, forensic psychiatrists testifying in cases of claimed physical disability do not answer the question, "Does this patient have a physical disability?" Instead, they rely on specialists in the areas of the patient's physical complaints—neurologists, neurosurgeons, orthopedists, gastroenterologists, cardiologists, and so forth. This is so because psychiatrists, like other specialists, have lost some of the skills they learned in medical school that are unrelated to their specialty. Even if they retained those skills, they would have to rely on other specialists because the law establishes "turfs," or territories, among specialties (Almy, 1983–1984). When a forensic specialist comments within the realm of expertise of another specialty, the lawyer whose client is disadvantaged by the comment will soon inquire of the intruding specialist, "Doctor, did you take a residency in that specialty? Do you have Boards in that specialty? For all practical purposes, those questions disqualify the specialist who answers no. Therefore, lawyers retain only those physicians who have all of the qualifications for the specialty, except when they cannot find a qualified specialist who is willing to present an opinion favorable to them (Blinder, 1982).

### Contrasting Treating Roles

Forensic psychiatry differs from clinical psychiatry in that prudent psychiatrists avoid treating patients with whom they have had a forensic role. Clinical psychiatrists strive to help their patients and are loath to testify to anything that would disadvantage their therapeutic relationship. Counter-transference affects their ability to testify objectively. Therefore, forensic psychiatrists who treat patients whom they have evaluated or evaluate patients whom they have just treated will have role conflicts.

These and other characteristics of the roles of forensic medical specialists suggest that they live and work in a different medical culture from that of ordinary health professionals. Except that forensic medicine is practiced by persons with similar education and general experience and that the forensic specialist is frequently a practicing physician who diagnoses and treats patients, forensic medicine and the rest of medical practice have little in common (Lasky, 1980).

## The Doctor-Patient Relationship

Physicians not engaged in forensic activity operate approximately in accordance with Talcott Parsons's (1951) description of doctor and patient roles. During most health care contacts there are no disputes. Patients come in voicing complaints; physicians take histories, examine the patients, order laboratory or X-ray studies, and collate the data. Physicians then inform their patients about the results and prescribe necessary treatment. Patients comply.

The basic assumption in these interactions—one without which the system could not function—is that patients are telling the truth as they know it, that they believe it is to their advantage to be healthy and unimpaired. While there are many departures from this model of interaction—disagreement, "doctor shopping," second opinions, and noncompliance—the basic theme is cooperation: both the patient and the physician are on the same side.

Doctor-patient roles are different in forensic medicine. Patients and their lawyers look for physicians who will declare them ill and disabled while the opposing counsel hope their specialists will find the patients less ill and more functional. While many patients worry that their physicians may not be competent, they do not usually question their honesty and integrity. In forensic medicine, patients frequently voice their suspicions, and even convictions, that specialists retained by the opposing side or even by their own lawyers have "sold out" to the employer or to the insurance company (Rothstein, 1984). Applicants' concerns about the bias of physicians can hinder open communication and the acquisition of data necessary to render a comprehensive and valid opinion.

## Questions of Causation

Workers' compensation statutes mandate benefits only if an injury arises out of employment or in the course of employment. Therefore, physicians must give opinions about what caused the injured worker's condition. The legal approach to causation differs from the medical approach. Within the legal approach, the defendant's attorney and the applicant's attorney offer differing viewpoints that favor their client's position. Attorneys and judges want to know if an identifiable event—incident, exposure, stress—was the "proximate cause" of the injury, either de novo or by "aggravating" a preexisting condition, thereby resulting in earlier or greater impairment than would ordinarily have occurred (Danner & Sagal, 1977).

In clinical work, competent physicians rely on a scientific approach to causation; that is, objective and verifiable diagnostic methods. In the forensic sphere, causation must be attributed on the basis of the best evidence available, with "reasonable medical probability." The specialist is forced to make a choice between alternative causes (Rothstein, 1984). Physicians and attorneys may disagree about the relative weight assigned a claimant's preexisting physical and psychological state compared with the work event alleged to be the proximate cause of the medical condition. Physicians and defense attorneys place great weight on a preexisting disorder. Attorneys and physicians representing the worker seek to prove that work events were the "straw that broke the camel's back" and therefore the cause of the injury. They seek to minimize preexisting and present causes and factors unrelated to work.

## Determining Presence of Disorders

In forensic medicine, the physicians cannot adhere to the usual initial hypothesis in diagnosing illness, "Where there's smoke, there's fire." Instead, they begin with the hypothesis, "Where there's smoke, there might be a smoke bomb." Under these circumstances, physicians determine the presence of disease or disorder by comparing objective signs, symptoms, and activity patterns with known physiological parameters.

For example, physicians know that total body pain or "wandering pain" is inconsistent with most physically caused syndromes. Claims of dysfunction that are consistent with known disorders and that evolve in ways consistent with known patterns will suggest the presence of an organic disease. When pain patterns are consistent with organic disease, there is little argument about the presence of disease, although there still may be disagreement about the cause. When the patterns are inconsistent, we encounter the forensic conflict in which the differential diagnoses of physical disease, affective disorders, somatoform disorders, and malingering are debated.

**Forensic Epistemology**

In judging the merits of arguments, courts and juries apply different criteria from those of most physicians who are not in the forensic arena (Bazelton, 1974). If physicians want to convince their fellow physicians that bodily change or sensations are related to a cause, they must follow accepted scientific principles of cause and effect. Nothing less will convince medical scientists. The same logic is required to demonstrate the relationship of treatment to outcome. In contrast, forensic experts are chosen because they are willing to represent one side's viewpoint as scientific, as that of an authoritative segment of the mainstream of medicine (Blinder, 1982). The term forensic applies to the debating process. A debate coached by opposing lawyers is conducted by the medical experts, who try to convince the judge and/or jury that the plaintiff does or does not experience what he or she reports or that it was or was not a work-incurred injury or that the worker's complaints were or were not due to a physical or mental disorder.

**The Forensic Questions**

Depending on the nature of the issues, the forensic psychiatrist must evaluate the pain and suffering, the mental distress, and the psychiatric disability caused by any physical disability (Lasky, 1980; Griffith, 1983). In personal injury cases heard in civil courts, for example, pain, suffering, and mental distress are compensable. In Workers' Compensation cases, the psychiatric issues include the need for treatment, the degree of disability as it affects competitive employment in the open-labor market, and the need for rehabilitation, but *not* the issue of pain and suffering.

In most disputed claims, several specialists from one field, for example, orthopedics, examine the patient. The forensic psychiatrist must decide whose opinion to accept as the basis for determining the patient's physical diagnosis. Frequently, the forensic psychiatrist has been called in because the physical specialists retained by both sides cannot explain the pain on a physical basis or because they agree about the condition but disagree about how much disability is imposed by psychiatric reactions to physical symptoms.

Forensic questions include the following (Mussoff, 1981):

- In a case where there is no objective physical basis for an applicant's symptoms, what psychiatric explanations and diagnoses can be made?
- If psychiatric diagnoses apply, how much suffering has been present and how disabled is the applicant by a mental disorder?
- Does the applicant have a mental condition that has caused an objectively determined physical disorder, such as a peptic ulcer, hypertension, coronary atherosclerosis, colitis, or the like?
- Is psychiatric treatment indicated to relieve the physical condition or to support the applicant psychologically? Is such treatment necessary to reverse a condition (e.g., depression) brought on by a physical disability?
- Which components of the psychiatric condition are reversible and which are irreversible?
- How will any irreversible psychiatric residuals affect the applicant's ability to:
    a. Work at a usual occupation,
    b. Work at any occupation,
    c. Care for himself or herself?

More general forensic questions include the following:

- Is the mental reaction a separate disorder from the physical disorder, and does it increase the level of disability already imposed by the physical disorder?
- What is a normal psychological reaction to a physical disability, and when should we consider a reaction excessive?
- If a predicted reaction is excessive or absent, should we consider that to be a sign of a mental disorder?

**Monetary Equivalence**

In both personal injury and workers' compensation forensic systems, all agreements and decisions must have monetary equivalents (Larson, 1978). In personal injury cases, pain and suffering must be translated into monetary compensation. For both these and workers' compensation cases, illness must be rated in dollar amounts calculated to reimburse victims for the impairment of function at home or on the job (Mussoff, 1981). Court decisions and settlements estimate future medical costs. Diagnosis, description of disability, indicated medical treatment, and rehabilitation are noted in order to provide payment. Some issues, for example, cosmetic changes and other deleterious effects on self-image, are often the most difficult to convert into money because although the loss of physical capacity to work may be minimal, the psychological impact from the loss is disproportionate to the physical change or impairment. The Workers' Compensation system addresses questions of disability, not questions of pain and suffering such as

- What price the loss of the tip of a finger when that loss does not interfere significantly with function?
- What price a scar about which the person obsesses?
- What price the loss of social ease?
- What price the loss of what the individual considered perfection?

## THE METHODS OF THE FORENSIC PSYCHIATRIST

### Records Review

The forensic psychiatrist's examination begins with a review of written records, including records from school, military service, civil and criminal courts, employment, and medical care, sometimes going back to the patient's childhood. Applicants and plaintiffs surrender their rights to the privacy of these records, including psychiatric records, especially when one side contends that they might be relevant to the contested issues. These records are studied in order to determine the patient's physical and mental state prior to the injury or illness in question. Even the most truthful patients have a tendency to forget or to minimize the severity of earlier symptoms, illnesses, or injuries. When the review is complete, the psychiatrist sometimes knows more about the patient's social and medical history than does the patient because the records contain information the patient has forgotten as well as laboratory data and the candid opinions of all the professionals who have dealt with the patient.

### Psychiatric Examination

The examination differs from that of the patient who comes for psychotherapy. The psychiatrist informs the patient of the purpose of the examination, discloses the referring source (employer's or applicant's attorney), and explains that nothing the applicant reports will be considered confidential. What is more, the psychiatrist does not know who, in addition to the referring source, will be reading the report. If the applicant agrees to proceed with the examination, the psychiatrist follows the procedures outlined in the standard texts (Kaplan & Sadock, 1980; MacKinnon, 1980a, 1980b). If the applicant's primary complaints are physical, special attention is paid to earlier physical symptoms, injuries, and disabilities. Forensic psychiatrists must answer specific questions, usually within a limited time frame. Therefore, both open- and closed-ended questions are asked, and the psychiatrist directs the course of the examination to obtain necessary information.

A thorough mental-status examination is done including testing of memory, orientation, comprehension, insight, judgment, the ability to calculate, and the ability to convert verbal directions with visual-motor work. The psychiatrist looks for evidence of pain or physical impairment and observes the subject's style of interacting.

At the end of the examination, the psychiatrist avoids taking on any aspect of the treating physician's role and does not discuss diagnosis or treatment recommendations, estimates of capacity to work, or the compensability of the condition. Nor does he or she comment on treatment given by other physicians.

Having reviewed medical and other records before examining the applicant, the psychiatrist can ask questions that might reconcile differences between the history presented by the applicant and the data in the records, and request additional records if necessary. After the data from the records and the examination are collated, the psychiatric examiner either recommends further psychological or physical studies or proceeds to answer the referring source's questions, giving the reasons for opinions and conclusions.

**Diagnostic Labels**

Physical disabilities can produce mental disorders. Injuries to the brain, for example, can produce alterations in thinking, feeling, and behavior. Pain, impaired body function, and cosmetic changes can in themselves result in depression or anxiety. They can cause other physical disorders as described in the third, revised edition of the *Diagnostic and Statistical Manual,* or DSM-III-R (American Psychiatric Association, 1987; Boehm, 1978), under the category, "Psychological Factors Affecting Physical Condition." Each case must be studied in order to determine the effects of the physical disability on the subject's mental state or the seeming lack of effect in some instances. Personality, impact on economic security, and impact on family are factors known to mediate the reactions to physical disability. The appropriate diagnostic label must be selected according to criteria in DSM-III-R. Categories relative to Mood Disorders and Adjustment Disorders are applied frequently.

When a person claims a physical disability in the absence of objective physical findings, we use the DSM-III-R labels of Histrionic Personality Disorder and the subgroups of the Somatoform Disorder category, especially Somatoform Pain Disorder and Conversion Disorder. Often the differential diagnosis is between one of the Somatoform Disorders and Malingering, the latter category being more frequently suspected than stated (Pollack et al., 1982). The category of Undifferentiated Somatoform Disorder is useful when people believe they are ill and their beliefs are supported by some physicians, but there is no evidence that they suffer a disease that could explain their symptoms (Brodsky, 1983a, 1983b). Examples are seen among those who believe they have been irreversibly harmed by exposure to chemicals in the absence of physical or laboratory abnormalities (Brodsky, 1983a).

## FORENSIC CHALLENGES

Generally, psychiatrists function in dyadic relationships with their patients and are challenged infrequently. When they are, however, they sometimes interpret challenges by patients as facets of transference rather than as valid intellectual differences. In contrast, in the medicolegal arena, psychiatrists' beliefs, methods, and judgments are questioned and challenged regularly (Bazelton, 1974; Boehm, 1978; Ziskin, 1975), and they are required to explain and justify their conclusions. For some, these challenges promote discovering areas of ignorance and rethinking beliefs and practices. Others treat these challenges as they treat disagreements with patients, and they interpret them as signs of the opponent's ignorance or hostility. Some respond to the challenge by developing unorthodox explanations, for example, that most mental illnesses are due to stress, toxic substances, allergies and impaired immune regulation, or vitamin and mineral deficits (Brodsky, 1983a, 1983b).

### Challenges of Testimony

During testimony, the lawyer asks the questions and the psychiatrist answers them. The questions are structured in terms appropriate to the lawyer's frame of reference. Frequently the psychiatrist cannot answer the question because of the way it is asked, and the lawyer does not know how to frame the question so that the psychiatrist can answer it. Sometimes the psychiatrist could restructure the question for the lawyer, but most experienced specialists answer the questions when they can and, when they cannot, say they do not understand the question or cannot answer it as asked.

Skilled lawyers sometimes use psychiatric interviewing techniques in order to elicit answers that cause the specialists to contradict themselves (Ziskin, 1975). Lawyers pretend they do not understand the responses and ask for clarifications. The specialists attempt to clarify but sometimes find that their efforts to elucidate only make them appear more confused, an outcome that discredits the testimony.

Frequently specialists are surprised when they are asked to comment on diagnoses or etiologic theories that had long been put to rest by medical researchers. They find that they are unprepared to discuss a diagnostic entity they do not believe exists or to discuss a relationship between symptoms and other external factors that they believe emanates from myths that have long been settled by research. In forensic circles, including those of workers' compensation adjudication, these issues are debated anew with each trial or hearing and with the assignment of new referees, namely, judges and juries. The fact that a medical issue might have been decided one way in a previous trial or hearing does not in itself increase the chances that it will be decided in the same way again (Bazelton, 1974; Blinder, 1982).

The workers' compensation litigating system has some special quirks; for example, psychiatrists or cardiologists who testify that stress does not cause coronary atherosclerosis might be disqualified because it has been held that they cannot testify about the intricacies of a relationship they do not believe exists. Legislatures, too, make decisions about causes of medical conditions. For example, the California legislature has decided that heart disease in a safety officer (police or fire departments) shall be considered work incurred and treated as if it were.

Usually, lawyers can find medical experts who will testify in favor of medical viewpoints that are far removed from those of physicians in the mainstream of medicine. One might ask why medical experts who testify contrary to scientific opinion or even contrary to prevailing lay opinion are not embarrassed or do not feel they have "sold" themselves. Some who testify firmly believe what they are saying and would refuse to testify to any other position. Others are prepared to develop arguments for or against a position in the manner of college debaters, who will defend positions contrary to their own beliefs and be proud of their skills if they carry the day (Almy, 1983–1984).

Another factor affecting the testimony of forensic specialists is sympathy for the underdog (Rothstein, 1984). The public tends to identify with the plaintiff in cases where the payer will be an insurance company or a government agency. The medical expert who slants data to help the "have-nots" against the "haves" is considered a "good guy," even by those who believe he or she is slanting that data. The expert then takes on some of the positive aura of the caring physician. We expect our physicians or other health care providers to be on the patient's side, for example, to produce a letter confirming that the patient was unable to work for a day or a week, even if that is not completely true. Bending the truth in the service of the patient is part of the "give" in our society. One public attitude is that no one will be hurt because the "deep pockets" of the insurance company are better able to afford the loss than are the individuals who claim injury.

**Challenge to Professional Integrity**

Forensic psychiatrists operate in an atmosphere in which there is increasing skepticism about their integrity. The press and, in some instances, the judiciary have suggested that one can find psychiatrists who will testify for or against any position and that there is a paucity of hard data to help the triers-of-fact decide on the merits of opposing testimonies (Ziskin, 1975). Some specialists remark that they can better tolerate skepticism about the validity of their views than they can assaults on their integrity.

**Quantifying Mental Distress and Disability**

Those involved in the process of granting benefits for mental distress and disability agree that it is sometimes difficult for both professionals and lay persons to quantify distress and disability (Griffith, 1983; Lasky, 1980). Some argue that psychiatrists and psychologists are no better qualified than lay persons to make such decisions. They criticize the reliance of courts on mental health professionals (Mussoff, 1981), suggesting that triers-of-fact are shirking the responsibilities that are theirs. They cite the writings of psychiatrists and behavioral scientists who consider mental illness a myth and a conspiracy, and they point to disagreements among mental health professionals who report and testify (Coleman, 1984). Some mental health professionals have prepared guidebooks that assist lawyers in cross-examining psychiatrists and psychologists in order to demonstrate that their opinions are based on data not susceptible to replication or to the usual tests of scientific validity. Nevertheless, the courts contend that they must rely on mental health professionals, not because they consider their testimony scientific, but because they are more reliable than lay persons (Bazelton, 1974). Therefore, forensic psychiatrists are asked to testify about the presence of mental disorder and its impact on the subject's ability to function, in a culture that challenges their qualifications to do so.

## Who Can Work: A Complex Question

Implicit in complicated cases of disputed disability are the broader questions:
- Who can work?
- Who should have the opportunity to work? (This question arises in cases of handicapped persons who are denied employment.)
- What physical or mental disorders or impairments excuse people from work and entitle them to be supported by others?

Some of these questions are answered in the labor marketplace itself. Employers need workers and select the best available. Choices may be limited by unions, by laws prohibiting discrimination in employment, by the current demands of the labor market, and by what the employer can pay and still operate profitably. During labor shortages, as in times of war, persons are hired who would be deemed unsuitable in peacetime. This is true in the military labor pool as well as in the civilian labor market (Brodsky, 1977b, 1984).

In making disability judgments, psychiatrists are asked to determine who can work and who will be able to keep a job after obtaining it, as if (1) The demands of the present labor market were not an important factor, (2) The effects of age on desirability as an employee were miniscule (Brodsky, 1971), (3) The prospective worker's motivation played no role in finding and keeping a job.

In reports written by psychiatrists addressing the issue of psychiatric disability, we find assumptions based as much on social values as on medical facts (Brodsky, 1977b; Ruesch & Brodsky, 1968).
- People should not have to work at jobs they no longer like or no longer find gratifying; "burnout" is a disability.
- They should not have to work at jobs they feel are dangerous.
- Being depressed is a disability in most jobs.
- People should not have to work at a job in which they are physically uncomfortable in any way.

- Mental disorder is a fiction, a myth, and the symptoms are consciously controllable.
- People who say they cannot work are lazy or unmotivated; any motivated person can work.

The realities of the workplace challenge these assumptions. We see those who are handicapped insist that they can work and demand equal opportunities. We see people in the workplace who have painful and life-threatening diseases, for example, cancer, ischemic heart disease, Parkinson's disease, and rheumatoid arthritis. We see some who leave the workplace at the first sign of discomfort, while others resist leaving in spite of pain and fatigue. Similarly, we see workers who have mental disorders of all kinds doing their jobs successfully, while coworkers are unaware that they have mental disturbances.

One striking case was that of a paranoid schizophrenic whose coworkers did not know that for years he had been hearing voices. They became upset, however, when he started answering the voices, and he was fired then. He sued, claiming that his mental state had not interfered with his ability to do his work and therefore should not disqualify him, and he was reinstated. Similarly, a fire fighter resisted retirement in spite of two back operations and finally was forcibly retired after he had a myocardial infarction. The differences between those who are markedly impaired but continue working and those who are unimpaired but claim disability are explainable in terms of personality, motivation, and self-image, among other factors. All these factors play a role in cases where individuals identify themselves in terms of their work roles and feel as if they will have no identity without their jobs.

## FORENSIC REHABILITATION ISSUES

Workers' compensation laws mandate rehabilitation, but disputes arise because there is no directive in those laws about the kinds of jobs for which impaired workers should be rehabilitated. Both the need for rehabilitation and the nature of the rehabilitation program are disputed. For example, an electrician wants to become an electronics engineer, while the employer or insurance company urges a less expensive program of training as a small-appliance worker. Some workers, unable to continue at their regular jobs because of a work-incurred injury, feel that they should be trained for higher-level jobs rather than jobs with lower salaries and lower status.

Some believe that as compensation for their physical losses, they should be given a second chance to achieve what they did not during their preinjury education. Further conflict occurs when the employer or the insurance company believes that the injured worker could be rehabilitated, but the worker claims he or she could not.

Psychiatrists are asked to comment about the following factors affecting the success of rehabilitation:

- Subjects' preference for the disability status over the job for which they could be trained.
- The monetary advantages and disadvantages of being rehabilitated.
- Feelings about work.
- Educational level and language ability, especially to read, write, and speak English.
- The capacity to learn new skills.
- Personality structure and adaptability, the capacity to get along and work with others, and the capacity to subordinate oneself to others as a student or trainee. Significant paranoid, passive-aggressive, or obsessive-compulsive traits can defeat the rehabilitation effort. Employers are more likely to tolerate the personality problems of long-term employees than to accept them in prospective employees who are already stigmatized by their need for rehabilitation.
- A pattern of substance abuse.

In spite of these and other problems, the rehabilitation mandate has served motivated workers well by restoring earning capacity, work status, and self-respect. It has served the community by converting dependent employees into active contributors to society's productivity.

## EFFECTS OF LITIGATION ON THE INDIVIDUAL

Litigation tends to fix patients' symptoms and their perceptions of disability. The effects of traumatically induced physical and psychological injuries, which usually diminish or disappear with time or treatment, persist even so, if the workers are involved in litigation. Some have attributed this delayed recovery and nonrecovery to greed, character weakness, or both. Although these factors may be present, other interpersonal elements exert equal force. The person who claims disability is being challenged. He or she can either give up the claim, thereby suffering a loss of self-esteem or a loss of the respect of family and friends, or seek to prove the statement of disability. The latter course, however, may make it more difficult to relinquish the signs and symptoms of disability, even after healing has occurred (Brodsky, 1984).

In disputed disability, the patient's credibility is suspect not only to adversaries but also to friends and family, who by their actions testify to their beliefs or their suspicions regarding the patient (Brodsky, 1984). Frequently the patient is teased by friends and family members, who hint that they believe his or her behavior regarding the disability is unnecessary. The patient may find it difficult to distinguish friend from foe and may feel isolated. The patient commits to a lawyer who will receive little or no payment if there is no disability to be compensated; this further reinforces the trend.

Many subjects who are plaintiffs in disability suits regret having filed their suits after they learn that they will have to go through multiple medical examinations, be questioned in depositions, and surrender their privacy by having medical, school, and work records opened to scrutiny. Some become suspicious of others and fear that they are being followed or that their phones are tapped. These persons may become more defensive and more forceful in asserting the authenticity of their disabilities.

For some, their claims become their lives. They abandon their former interests, activities, and associations. They talk about their claims most of the time. They write numerous lengthy letters to public officials, judges, appeals bodies, and their own attorneys. If they reach a point where attorneys are no longer willing to represent them, they become their own lawyers, acting "in pro per" and preparing their own briefs. They study law informally and learn and use the proper legal forms. Some involve their spouses, and some spouses direct the case, even when the claimant is represented by lawyers. Not infrequently, the legal claim becomes a family matter that binds the members together more strongly than ever before. Almost invariably, litigation itself causes the subjects some distress, regardless of the merits of their disputed disability claims.

## POSTLITIGATION STAGE

Remarkably little attention has been paid to the postlitigation stage of the applicant's life. In most instances, lawyers have no continuing financial interest after litigation is terminated. The insurance company pays on all of the bills and awards, accepts the loss, and is no longer concerned with whether the awards and payments were inadequate or excessive for the level of disability.

Postlitigation follow-up studies describe a variety of outcomes, but it is evident that many people do not recover simply because (or when) litigation is ended (Balla & Moraitis, 1970). Postlitigation behavior can be studied in workers' compensation applicants who have histories of previous injuries, litigation, and settlements and who, after finding new jobs, suffer new injuries or aggravation of their old symptoms. The behavior of workers who have new symptoms associated with the original injury and who claim that the original award for disability and medical treatment was inadequate, has also been studied. Another source of data is the death claim, filed by the worker's dependents, alleging that the death or suicide was the result of a work-incurred injury that had already been litigated and resolved within the Workers' Compensation courts (Brodsky, 1977a).

A review of such cases shows that many applicants are satisfied with the outcome of litigation and use their awards to restructure their lives to the extent possible. Others squander any lump-sum payment awarded and become depressed when they can find no satisfactory means to support themselves. Some become morose when the excitement of the litigation is over, and they can find nothing to replace it with.

Litigation can cement family relationships and provide a common interest and activity for family member. The uninjured spouse may become the director and facilitator of litigation and frequently defeats settlement attempts because he or she fears the loss of position. Some families break up when the common interest and effort in litigation ends and they are faced with the realities of their occupational and economic problems without their former hopes of remedying those through successful litigation. Some families break up under the initial stress of coping with the economic and social changes resulting from the disability.

## SUMMARY

Law and psychiatry operate from two inherently different perspectives, and the forensic psychiatrist needs to understand the special culture and the differing conceptualizations and models of reasoning of both fields in order to apply psychiatry to the legal issues involved in the evaluation of the disability claims of workers' compensation applicants.

Medicolegal psychiatry concerns itself with a particular aspect of the interface between social pressures and personal dynamics and can be commended to those seeking new experiences. It is a looking-glass land in which rare syndromes become commonplace, treatments do not have their anticipated effects, and illnesses flow along courses which no ordinary cartographer could anticipate. One may also learn something of the motivations and identifications of doctors, the true meaning of the words "secondary gain," and that there are many things in heaven and earth which do not emerge in a psychiatrist interview. (Ellard, 1970, 349)

This chapter has discussed some of the interactions of psychiatry and law related to the determination of disability and the resolution of disputes about disability. The legal culture is dominant in this relationship and determines the logic and the epistemology by which disputes about disability are decided. Psychiatrists experience "culture conflict" because they are pressed to accept the rules, roles, values, and ethics of the legal system and its professionals.

Forensic psychiatrists resolve these culture conflicts in different ways. Some avoid professional situations that would put them into this kind of conflict, while others enjoy the challenge of fitting together the pieces of these two systems. Some enjoy the license to be adversarial, while others are shaken by challenges to their professional authority.

The process of resolving disability disputes illustrates the more general elaboration of dispute resolution in our society. A "no fault" system for providing benefits for injured workers has evolved into a specialized field for lawyers and physicians, a field in which the medicolegal costs contribute substantially to the expense of the system (Witt, 1985).

## REFERENCES

Almy, T.B. (1983–1984). Psychiatric testimony: Controlling the "ultimate wizardry" in personal injury actions. *The Forum, 19,* 233–267.

American Psychiatric Association. (1987). *Diagnostic and statistical manual of mental disorders* (3d ed., rev.). Washington, DC: American Psychiatric Association.

Balla, J.I. & Moraitis, S. (1970). Knights in armor: A follow-up study of injuries after legal settlement. *Medical Journal of Australia, 2,* 355–361.

Bazelton, D.L. (1974). Psychiatrists and the adversary process. *Scientific American, 230,* 18–23.

Behan, R.C. & Hirschfeld A.H. (1963). The accident process. I, II. *Journal of the American Medical Association, 186,* 193, 300.

Behan, R.C. & Hirschfield, A.H. (1966). The accident process. III. *Journal of the American Medical Association, 197,* 85.

Blinder, M. (1982). *Psychiatry in the everyday practice of law* (2d ed.). San Francisco: Bancroft Whitney.

Boehm, D.O. (1978). DSM-III and

the legal system. *Bulletin of the American Academy of Psychiatric Law, 6,* 31–35.

Brodsky, C.M. (1971). Compensation illness as a retirement channel. *Journal of the American Geriatric Society, 19,* 51–60.

Brodsky, C.M. (1977a). Suicide attributed to work. *Suicide and Life-Threatening Behavior, 7,* 216–229.

Brodsky, C.M. (1977b). The genesis of a problem population. In P. Ostwald (Ed.), *Communication and social interaction.* New York: Grune & Stratton, 119–132.

Brodsky, C.M. (1983a). Allergic to everything: A medical subculture. *Psychosomatics, 24,* 731–742.

Brodsky, C.M. (1983b). Somatoform disorders in the workplace. *Journal of Occupational Medicine, 25,* 459–464.

Brodsky, C.M. (1984). Sociocultural and interactional influences on somatization. *Psychosomatics, 25,* 673–680.

Coleman, L. (1984). *Reign of error: Psychiatry, authority, and the law.* Boston: Beacon.

Danner, D. & Sagall, E.L. (1977). Medicolegal causation: A source of professional misunderstanding. *American Journal of Law and Medicine, 3,* 303–308.

Ellard, J. (1970). Psychological reactions to compensable injury. *Medical Journal of Australia, 2,* 349–355.

Galanter, M. (1983). Reading the landscape of disputes: What we know and don't know (and think we know) about our allegedly contentious and litigious society. *UCLA Law Review, 31,* 4–71.

Gorman, W.F. (1983). Are there impartial expert psychiatric witnesses? *Bulletin of the American Academy of Psychiatry and Law, 11,* 379–382.

Griffith, L.E., Jr. (1983). Recovery under workers' compensation statute for mental disorder caused by work-connected stress without physical cause or result. In *Shepard's causes of action* (Vol. 1). Colorado Springs: Shepard's/McGraw-Hill. 75–149.

Harter, P.J. (1983–1984). Dispute resolution and administrative law: The history, needs, and future of a complex relationship. *Villanova Law Review, 29,* 1393–1419.

Heiman, E.M. & Shanfield, S.B. (1978). Psychiatric disability assessment: Clarification of problems. *Comprehensive Psychiatry, 19,* 449–454.

Kaplan, H.I. & Sadock, B.J. (1980). Psychiatric report. In H.I. Kaplan, A.M. Freedman & B.J. Sadock (Eds.), *Comprehensive textbook of psychiatry/III* (Vol. 1). Baltimore: Williams & Wilkins, 899–905.

Kapp, M.B. (1985). Medicine and law. A symbiotic relationship? *American Journal of Medicine, 78,* 903–907.

Larson, A. (1978). *The law of workers' compensation.* New York: Matthew Bender.

Lasky, H. (1980). Psychiatry and California workers' compensation laws: A threat or a challenge? *California Western Law Review, 17,* 15–38.

MacKinnon, R.A. (1980a). Diagnosis and psychiatry: Examination of the psychiatric patient. In H.I. Kaplan, A.M. Freedman & B.J. Sadock (Eds.), *Comprehensive textbook of psychiatry/III* (Vol. 1). Baltimore: Williams & Wilkins, 899–905.

MacKinnon, R.A. (1980b). Psychiatric history and mental status examination. In H.I. Kaplan, A.M. Freedman & B.J. Sadock (Eds.), *Comprehensive text-*

book of psychiatry/III (Vol. 1). Baltimore: Williams & Wilkins, 906–920.

McMahon, E.A. (1983). Forensic issues in clinical neuropsychology. In C.J. Golden & P.J. Vicente (Eds.), *Foundations of clinical neuropsychology.* New York: Plenum, 401–427.

Mussoff, J. (1981). Determining the compensability of mental disabilities under workers' compensation. *Southern California Law Review, 55,* 193–253.

Parsons, T. (1951). *The social system.* Glencoe, Ill.: Free Press.

Pollack, S., Gross, B.H. & Weinberger, L.E. (1982). Dimensions of malingering. In B.H. Gross & L.E. Weinberger (Eds.), *The mental health professional and the legal system.* San Francisco: Jossey-Bass.

Rothstein, M.A. (1984). Legal issues in the medical assessment of physical impairment by third-party physicians. *Journal of Legal Medicine, 5,* 503–548.

Ruesch, J. & Brodsky, C.M. (1968). The concept of social disability. *Archives of General Psychiatry, 19,* 394–403.

Witt, M.S. (1985). *California Workers' Compensation Reporter, 13*(4), 84 passim.

Ziskin, J. (1975). *Coping with psychiatric and psychological testimony* (2d ed.). Beverly Hills, Calif.: Law and Psychology Press.

Zusman, J. & Simon, J. (1983). Differences in repeated psychiatric examinations of litigants to a lawsuit. *American Journal of Psychiatry, 140,* 1300–1304.

# 13
# The Poverty of Justice: Ethics and the Physically Disabled

Robert M. Nelson

The Rehabilitation Act of 1973 was designed to end discrimination on the basis of physical disability. Section 504 of the act granted sweeping legal rights for handicapped individuals (Hahn, 1983, 37).[1] Recently this section and its role in the so-called Baby Doe regulations have been the focus of much controversy involving the medical and surgical treatment of severely disabled newborns (Annas, 1986; Lyon, 1985; Shelp, 1986). The United States Supreme Court has held that section 504 does not apply to individual treatment decisions in which the disability itself is a factor in determining the benefit derived from the intervention (*Bowen v. American Hospital Association*, 1986). This reasoning closely follows the conclusion of the Second Circuit United States District Court of Appeals. In that conclusion, the phrase "otherwise qualified" indicates that section 504 applies to disabled persons who would benefit *regardless* of their disabilities, as opposed to those whose disabilities are themselves the focus of intervention (*United States v. University Hospital*, 1984). In addition, the court found that the monitoring and control of individual treatment decisions should fall under state protective functions for child abuse and neglect, which stand outside of the nondiscrimination mandate of section 504.[2]

The central issue, however, remains: Is the presence and extent of a physical impairment sufficient reason for not treating an infant who otherwise would be treated? A classic example is of a severely paraplegic infant who was born with a high myelomeningocele and who showed no evidence yet of central neurological damage. With aggressive surgical management of the hydrocephalus, which inevitably develops after closure of a myelomeningocele, the infant stood a good chance of normal cognitive development. However, on the basis of the infant's paraplegia, the initial selection criteria published in the early 1970s excluded the child from treatment on the argument that subsequent suffering would be even greater in the presence of normal intelligence (Lorber, 1971, 1976; Weir, 1984). Furthermore, a rough formulation of the infant's expected quality of life was developed, taking into account the infant's medical condition as well as social and economic contributions from family and society (Shaw, 1977, 1988).

On the basis of this kind of formulation, some infants with severe myelomeningocele are not treated out of concern that the infant's family would be unable to provide adequate care (Gross et al., 1983). Decisions to allocate needed medical and surgical treatment according to the availability of family and social resources have been criticized as an unjust discrimination, which violates an infant's rights (Ramsey, 1978, Venes, 1984; Nelson, 1980). However, such decisions have been defended when the costs and burden of care are sufficiently great to "defeat customary duties of beneficence" (Shelp, 1986; Duff, 1981). Both sides agree that the financial and psychological support made available by society to these disabled infants and their families is often woefully inadequate.

The plight of these infants and their families has generated considerable discussion concerning the ethical and social issues surrounding medical care of the disabled. This discussion may be viewed as part of the broader issue of society's treatment of disabled persons. Does the Court's decision that section 504 applies only in situations where the disability is not relevant to the benefit in question imply that an individual may be denied employment if the disability undercuts productivity? Or should the decision be more narrowly interpreted as simply excluding section 504 from impinging on individual medical decisions? Do considerations of justice adequately address our treatment of the disabled or at least adequately establish the extent of government intervention and enforcement?

The descriptive words we use often convey moral values, such as our choice of either "fetus" or "infant" in discussing abortion. Recognizing the value-laden aspect of descriptive concepts, those who write about physically disabled persons prefer to use the term *disability* rather than *impairment* or *handicap*. Whereas the concept of impairment refers to a strictly functional limitation or to a loss of a bodily organ, the concept of disability relates this limitation or loss to the performance of expected social roles or tasks while avoiding the negative connotations of the concept of handicap. Handicap is a more obviously social construct, which may include external barriers or disadvantages imposed by society upon the disabled (Acton, 1982; Roth, 1983; Hahn, 1983). This concern with language is linked to a more general criticism about the overly individualistic focus of our social policy toward the disabled. We have tended to view physical disabilities primarily as an individual medical problem; we seek to eliminate, compensate, or reduce the particular impairment while we fail to address the social, political, and environmental causes of handicapped lives. A conscious effort to adopt and use the term disability may serve as a corrective by incorporating an awareness of social roles and tasks (Acton, 1982; Hahn, 1983).

The political struggle for social equality on the part of disabled persons has benefited, nevertheless, from this focus on the individual. The civil rights movement was the first step in the development of rights for the disabled, starting with *Brown v. Topeka Board of Education* in 1954 and leading through the Rehabilitation Act of 1973 (including Section 504) and the Education for All Handicapped Children Act of 1975 (Roth, 1983). The language of rights oriented toward the protection of individual *liberty* is based on a "severe individualism" (Will, 1986). However, to attempt to improve the welfare of the disabled through an appeal to individual rights that is based on *equality* has within it an inherent tension. This is illustrated by a problem with Public Law 94–142, the Education for All Handicapped Children Act of 1975, in which an apparent conflict exists between the goal of integration and the need for and assumptions of special education by way of appropriate instruction and due process (Hahn, 1983). At the same time that one asserts individual moral equality and rights of full participation by denying the significance of the disability, one is making an appeal for special rights and treatment on the basis of being disabled. Being disabled must be seen as ethically relevant in order to be the basis of a social entitlement (Will, 1986).

The ambiguity concerning the ethical relevance of being disabled stems from a failure to specify the context or purposes for which the judgment of equality is being made. The accepted principle for the fair or just distribution of social resources states that one should treat equals equally and unequals unequally according to their relevant differences. This general principle of distributive justice is incomplete without criteria for determining the relevance of similarities and differences among persons (Benn, 1967; Feinberg, 1978). For example, as discussed earlier, the United States Supreme Court held that the phrase "otherwise qualified" limited the applicability of section 504 to those circumstances, such as education and employment, in which the disability is irrelevant to the benefit in question. The presence or absence of a physical disability is clearly relevant to the provision of individual medical treatment and thus, in the opinion of the Court, lies outside the purview of section 504.

At issue then is the moral justification of defining differences between persons in order to provide services and benefits designed to yield equal opportunity.[3] Are there differences between persons that should not be a factor in determining their share of available goods, that is, their ability to lead a fulfilled life? Clearly there are some social goods we are willing to allocate according to the existing distribution of physical ability, such as accolades for athletic skill or great beauty. We do not accept, however, that the ability to board a bus or climb stairs should limit one's opportunity to take out a library book or get a job. The concept of equal opportunity is thus vague and ambiguous until we specify the agent that should have a chance to attain a goal without the hindrance of a particular obstacle. The notion of equality derives from this specification in stating that the persons involved are identical according to the measure stipulated as applicable to the given opportunity (Weston, 1985).

The complexity of equal opportunity and the sense of justice that it is intended to reflect thus rest on the diversity of our judgments concerning various goals, pertinent obstacles, and the necessary effort or qualifications to reach those goals. There is no simple formula for determining the desirability of equal opportunity for a particular goal and thus no simple determination of the moral requirements of equality and justice. Much of the confusion that prevails concerning the practical demands of equal opportunity would dissipate before a clearer understanding of the specific obstacles and intended goals. Recall the tension mentioned earlier between integration and special education: in the absence of educational goals, it is impossible to decide which approach would enable the disabled person to most fully participate within society. Whether the presence of a disability requires special treatment in order for the disabled person to have an equal opportunity cannot be answered in general. An answer requires a specification of the nature and goals of the opportunity in question as well as the obstacles to overcome.

Unjust discrimination in employment may be the major current issue facing the disability movement (Roth, 1983). Two aspects of employment discrimination can be identified: hiring and wages. Present medical and vocational rehabilitation focus on attaining employment, often without regard to wage discrimination or the cost of hiring to the employer. Though one study found that 35% of the wage differential between disabled and nondisabled workers is due to discrimination, the difficulty in directing policy is that the cause of the discrimination—the disability—may also reduce productivity. Should the disabled receive special treatment in the workplace to account for lower productivity? Should they only be given jobs where productivity is unaffected by the disability? Statutory and administrative prohibitions against discrimination have tended to emphasize equity or the need for special treatment rather than efficiency, thus requiring the employer to absorb the costs of hiring, training, and possible decreased productivity. An alternative approach is to recognize the trade-off between equity and efficiency and thus distribute the costs of achieving equal opportunity across all members of society (Johnson & Lambrinos, 1983). This approach is consistent with Sidgewick's (1907) distinction between "ideal" and "conservative" justice in which the transactional costs in moving from the present to the presumed ideal and just state are taken into consideration.

The failure to deal adequately with these transactional costs in providing for the social integration and economic participation of the disabled may work against the claimed cost-effectiveness of independent living. Presumably, a disabled person would require either modification of the workplace or special tools to compensate for his or her disability. The disabled person's productivity would be either equal to or less than a qualified nondisabled person.[4] Given the cost of the modification or special tool, the overall cost to the business per unit of productive work would be greater for the disabled person unless such transactional costs were shared more broadly within society. In essence, the savings implied by arguments for the greater cost-effectiveness of independent living for the disabled need to be fairly distributed.

In the absence of a social policy providing for such a sharing of burdens and benefits, is there any guarantee that the tension between equity and efficiency within the workplace will be resolved in favor of equity? As mentioned earlier, the simple appeal to equal individual rights may neglect the relevance of disability to individual productivity and the increased employer costs of hiring, training, and necessary special facilities. As a result, these constraints on efficiency will compel many businesses to provide minimal protection for the rights of the disabled and to hire only to the extent necessary for protection from litigation (Johnson & Lambrinos, 1983). More generally, the adversarial process involved in the establishment and enforcement of individual rights risks the loss of spontaneous sympathy for the plight of the disabled (Will, 1986). To advocate a conceptual shift from an individual to a social point of view is not without similar risk. Calling for the development of a political organization among the disabled for the purpose of overall sociopolitical recovery risks the loss of public sympathy, which has been predicated on the apolitical nature of the disabled as sick and thus deserving (Roth, 1983).

Noting the ugliness that can follow from a simplistic clash of individual rights that is based on a narrow view of autonomy, such as in the Elizabeth Bouvia case, Eric Cassell (1986) has suggested the appreciation of an "aesthetic" dimension in which life is seen as a "work of art." Each individual is in the process of being shaped by a series of choices made in harmony with the person that individual was or is becoming. The individual is not viewed solely as a discrete or separate entity in time; thus, autonomy serves a useful function by promoting an understanding of the role each person has in creating, over time, the texture and design of his or her own life. This same appreciation for the place of an individual within the community informs the writing of George Will (1986) though he extends the discussion beyond that of rights and autonomy. The welfare of the disabled depends on the virtue of benevolence within the community based on a sense of common vulnerability and neediness. Class action litigation and other assertions of group rights can impede the emergence of benevolence in the community.

A de-emphasis on justice combined with an appeal to community benevolence, as Will suggests, appears to undercut the possibility of organized political action by the disabled. Must the disabled resign themselves to community voluntarism or individual appeals to charity in apparent rejection of the successful civil rights strategy of the past 15 years? Such a resignation assumes that there can be no enforceable duties of charity. Yet, given that morality is fundamentally concerned with the well-being of others, an argument can be made for the existence of duties of charity, for example, a duty to provide an easy or "reasonable" rescue or aid. In addition, enforcement is sometimes necessary and justified in order to secure contributions to collective goods, regardless of an individual's moral right to the good in question. The moral justification for enforcing contribution does not necessarily presuppose either legal or moral rights to the collective good (Buchanan, 1987). Such an argument in favor of a duty of charity would be complex and controversial, but it is not excluded from the outset. The practical establishment of such a duty of charity would involve a complex political process of forming a consensus of community values concerning our moral obligation toward the disabled. Legislation rather than litigation would be the programmatic focus of advocates for the disabled (Will, 1986).

The appeal to charity may wear thin, however, in the face of a scarcity of available resources. Philosophers have long considered the presence of "moderate" scarcity as the condition under which justice becomes necessary

(Benn, 1967; Feinberg, 1978). Justice then becomes a way of adjudicating between the various claims on these resources. Consider a recent discussion of a possible "duty to die" on the part of the elderly. Given the apparent scarcity of medical resources and the great needs of elderly disabled persons, it is argued that the direct termination of these lives more nearly achieves the requirements of justice for society as a whole than does the denial of medical treatment to the disabled elderly. This position assumes that the scarcity is real and that a subsequent redistribution of medical resources to the young would in fact occur. It also recognizes that we may choose to redistribute our resources in the face of such a consequence, thus eliminating the scarcity (Battin, 1987; Daniels, 1985). Why do we recoil from such a conclusion? What prevents us from redistributing our resources in this fashion if a just efficiency argues in favor of such a policy (Wolfensberger, 1981)?[5]

The conclusion that the direct termination of the lives of the elderly disabled rather than the denial of medical treatment more nearly achieves the requirements of justice, given scarce resources, demonstrates the moral poverty of justice alone in determining social policy (Battin, 1987). It also reveals the danger of discussing the practical implications of scarce resources in the absence of an overall policy governing the ordering of social, economic, and political goods. Given the lack of a coherent health policy, it is necessary and appropriate for disabled individuals and their families to seek the resources they need for full and fruitful participation within society. To do otherwise would be to opt prematurely out of a political process that should have as an end point the formulation of a set of national health priorities. The burden of scarce resources ought to be shared equally by all members of society.

Within the public arena, moral debate concerning the treatment of disabled persons has long been dominated by considerations of justice and individual rights. Though much has been accomplished, the perspective of justice has distorted our moral responsibility toward the disabled. Justice without compassion and caring can isolate persons and subsequently destroy both the individual and the community. The achievement of a humane society in which each one of us experiences an equality of dignity and well-being requires an ethic of mutual care and responsibility, which incorporates and moves beyond the requirements of justice (Flanagan & Jackson, 1987). An overemphasis on justice risks rejecting the unifying principle of caring; however, an overemphasis on caring risks losing the sense of justice, which is the common public form in which caring is actualized (Tillich, 1954). The moral treatment of the disabled person within society requires tempering justice with compassion, for otherwise the alleged scarcity of resources would serve to excuse overt discrimination and even death by neglect or active termination.

## NOTES

1. Section 504 states that "no otherwise qualified handicapped individual in the United States . . . shall, solely by reason of his handicap, be excluded from participation in, or be denied the benefits of, or be subjected to discrimination under any program of activity receiving Federal financial assistance."

2. In the meantime, Congress had incorporated restrictions on the withholding of medically indicated treatment from disabled infants in the 1984 Child Abuse Amendments, Public Law 98–457. These amendments serve as the current "Baby Doe regulations" in that they define as medical neglect the failure to provide indicated treatment.

3. The predominant interpretation, at least in the United States, of the equal treatment required by the principle of justice is that of an equality of opportunity. Rather than distributing to each person an equal amount of goods, we seek to create a social structure in which each person can compete on an equal footing for a variable share of goods. Certain goods necessary to the basic social structure may be called "primary" goods and thus distributed in a different fashion than achieved "secondary" goods.

4. This discussion is meant as a generalization and ignores the possibility of a gifted disabled person being able to outperform a nondisabled person. While this is always a possibility, it is more often the exception rather than the rule. The discussion also assumes that the purported cost-effectiveness of independent living includes such transactional costs as part of the overall economic analysis. Otherwise, the argument holds out the promise of savings in one sphere while incurring more costs in another.

5. Within Nazi Germany, the extermination of the handicapped was seen as the logical outcome of a policy of efficiency, that is, the appropriate utilization of social resources.

## REFERENCES

Acton, N. (1982). The world's response to disability: Evolution of a philosophy. *Archives of Physical Medicine and Rehabilitation, 63*, 145–149.

Annas, G. (1986). Checkmating the Baby Doe regulations. *Hastings Center Report, 16*, 29–31.

Battin, M. (1987). Age rationing and the just distribution of health care: Is there a duty to die? *Ethics, 97*, 317–340.

Benn, S.I. (1967). Justice. In P. Edwards (Ed.), *The Encyclopedia of Philosophy* (Vol. 4), 298–302.

Bowen v. American Hospital Association, 54 LW 4579 (June 9, 1986).

Buchanan, A. (1987). Justice and charity. *Ethics, 97,* 558–575.

Cassell, E.J. (1984). Life as a work of art. *The Hastings Center Report, 14* 35–37.

Daniels, N. (1985). *Just health care.* Cambridge: Cambridge University Press.

Duff, R.S. (1981). Counseling families and deciding care of severely defective children: A way of coping with "medical Vietnam." *Pediatrics, 67,* 315–320.

Feinberg, J. (1978). Justice. In W.T. Reich (Ed.), *Encyclopedia of Bioethics* (Vol. 2), 802–811.

Flanagan, O. & Jackson, K. (1987). Justice, care and gender: The Kohlberg-Gilligan debate revisited. *Ethics, 97,* 622–637.

Gross, R. H., Cox, A., Tatyrek, R., Pollay M., & Barnes W., (1983). Early management and decision making for the treatment of myelomeningocele. *Pediatrics, 72,* 450–458.

Hahn, H. (1983). Paternalism and public policy. *Society, 21,* 36–46.

Johnson, W. G. & Lambrinos, J. (1983). Employment discrimination. *Society, 20,* 47–50.

Lorber, J. (1971). Results of treatment of myelomeningocele: An analysis of 524 selected cases, with special reference to possible selection for treatment. *Developmental Medicine and Child Neurology, 13,* 279–303.

Lorber, J. (1976). Ethical problems in the management of myelomeningocele and hydrocephalus. *Nursing Times, 72* (26 February), 5–8; and *Nursing Times, 72* (25 March), 9–11.

Lyon, J. (1985). *Playing God in the nursery.* New York: W. W. Norton

Nelson, R. (1980). Spina bifida: A medical and moral dilemma. M. D. Thesis, Yale University.

Ramsey, P. (1978). *Ethics at the edges of life.* New Haven: Yale University Press.

Roth, W. (1983). Handicap as a social construct. *Society, 20,* 56–61.

Shaw, A. (1977). Defining the quality of life. *The Hastings Center Report, 7,* 11.

Shaw, A. (1988). QL revisited. *The Hastings Center Report, 18,* 10–12.

Shelp, E. E. (1986). *Born to die? Deciding the fate of critically ill newborns.* New York. The Free Press.

Sidgewick, H. (1907). Justice. In *The methods of ethics* (7th ed.). New York: Dover.

Tillich, P. (1954). *Love, power and justice.* New York: Oxford University Press.

United States v. University Hospital, 729 E. 2d 144 (1984).

Venes, J. (1984). Care of infants with myelomeningocele (letter). *Pediatrics, 74,* 162–163.

Weir, R. (1984). *Selective nontreatment of handicapped newborns.* New York: Oxford University Press.

Westen, P. (1985). The concept of equal opportunity. *Ethics, 95,* 837–850.

Will, G.F. (1986). For the handicapped, rights but no welcome. *The Hastings Center Report, 16,* 5–8.

Wolfensberger, W. (1981). The extermination of handicapped people in World War II Germany. *Mental Retardation, 19,* 1–7.

# 14
# A History of The Independent Living Movement: A Founder's Perspective

*Edward V. Roberts*

Before the civil rights movement, black people had to go to the back of the bus to find seating. Many people with disabilities could not even get on the bus. In the 1970s the civil rights movement opened the way for disabled people to integrate into society, to gain control over their lives, to move out of institutions.

People with disabilities have needed a civil rights movement because of society's fundamental prejudices toward disability. Erroneous perceptions have dominated society's thinking and, therefore, public policy for centuries. Most prevalent is the perception that people with disabilities are nonproductive members of society who must be taken care of, either by families or institutions, during their entire life spans. Our language reflects these attitudes in such words as *invalid*, which as an adjective means "of no force," but as a noun means "one who is weak or infirm." Countless other words like *idiot, deaf and dumb, retard, freak, lame, loonie,* or *cripple* are prejudicial expressions.

## THE MEDICAL MODEL

Historically, people with disabilities have been seen as physically or mentally ill, as permanent clients of the "medical model" of lifetime care.

Though served by the medical profession, people with disabilities are also often devalued by it. The medical profession cannot cure the disabled, treatment of people with disabilities is often considered a burden because it provides low economic rewards and is therefore accorded low priority.

The medical model has also contributed to the low expectation society holds for the disabled and to a general acceptance that people with disabilities are nonproductive members of society. As is so often the case, low expectations become self-fulfilling. A further irony of the medical model is that, while medicine continues to keep more and more people alive even after traumatic accidents or life-threatening diseases, it offers little or no support for their return to the community as disabled citizens. Too often they become dependents of families or wards of institutions.

Clearly, people with disabilities represent a much larger group of society than has been previously imagined. Today, there are 32.6 million people with disabilities in the United States (National Center for Health Statistics, 1986). Maybe a better approach would be to think of all people as "temporarily able-bodied," since during our lifetimes most of us will have a disability severe enough to require a barrier-free environment, accessible transportation, and adapted housing.

## HISTORY OF DISCRIMINATION

Modern stereotypical perceptions about people with disabilities began with the industrial revolution. Industry needed able hands, strong backs, and stamina in workers. People with disabilities, which were often acquired through industrial accidents, diseases, or overwork, became both "excess baggage" for social Darwinists and objects of charity. Industrial societies in the 19th and 20th centuries produced enough capital to care for the disabled. "Care," however, implied isolation. People with even modest disabilities were often placed in charitable institutions and segregated from the rest of society. It was naively believed that these institutions could make their inmates more "normal." Instead, institutional life socially disabled whole generations of people.

Support for institutionalization came from many quarters. People wanted—and still want—institutions to relieve them of responsibility for the care of the disabled. Parents felt they needed institutions to care for their children when they, the parents, died. People with disabilities, afraid of what would happen to them without the support of their parents, accepted institutionalization. Fear of the future is still one of the strongest forces perpetuating institutional care and making people resist independent living.

Pseudoscientific theories were invented to justify institutionalization. For example, it was believed that mentally retarded persons should be separated from the general population so as not to pollute the gene pool and propagate a race of imbeciles. Not only was this theory erroneous, but it was also illogically applied, since parents who produced the mentally deficient offspring were never sterilized or prohibited from producing more offspring. Having a mentally retarded child imposed a stigma on parents. Yet, it was the children who were damaged; they were removed from the community and placed in segregated institutions where they would be out of sight and out of mind (Wolfensberger, 1977).

Charitable fundraising also stereotypes people with disabilities. In print, radio, and later television—especially in the telethon—an image was created of disabled people as weak, sick, or dying. The classic example is the March of Dimes, which propagated and perpetuated the worst images of people with disabilities. The March of Dimes poster child was the epitome of helplessness, chosen to arouse people's pity.

The guilt and fear engendered by the charitable approach has burdened the public and has created a major barrier to equal opportunity and full integration into the community. To this day people are afraid of men and women with disabilities. This approach has also specialized disabilities thereby leading to a fragmentation of the disabled community and its supporters. (The charities devoted to cerebral palsy have little common cause with charities devoted to the hearing impaired, for example.) Charities have spent a great deal of money perpetuating themselves and finding "the cure," while neglecting other important issues such as quality of life. Certainly, finding a cure is important but not more so than fostering meaningful lives for people who have the disability now, with no expectation for a cure in their lifetime. Waiting for a cure doesn't help people deal with their present situations or move ahead with their lives. The issue for them is the quality of their lives today. Lately, some charitable organizations have begun to recognize the importance of independent living.

The medical rehabilitation model emerged in the 1940s. It did not help to create a positive image of people with disabilities. The ideal case was that of a person going into the hospital flat on his or her back, working from this position to sitting and standing, and finally walking out of the hospital as a "normal person." For millions this would never happen; they considered themselves rehabilitative failures. Those who did not walk were sent home or to an institution and were perceived as sick for the rest of their lives.

Another example, from my own life, shows how the negative images of disabilities are transferred to families and friends of the disabled. After I contracted polio at the age of 14, my parents asked the doctor if I would live or die. He told them it would be best if I died because if I lived, I would be nothing but a vegetable for the rest of my life. I am now known the world over as the "artichoke"—prickly on the outside with a big heart on the inside. The independent living movement is bringing together the vegetables of the world!

The medical profession has contributed to the compartmentalization and dehumanization of people with disabilities. Many medical professionals (physical therapists, occupational therapists, social workers, and doctors) have looked only at the disability and have ignored the person as a whole. It was, and still is, difficult for a disabled person to get answers to such questions as: Can I have sex and a family? Will I be able to work? Where will I live? Professionals tend to limit themselves to their areas of expertise without using creativity or sensitivity about the disabled person's situation, when this perceptiveness might allow them to make appropriate suggestions. One of my therapists insisted that I learn to feed myself. Meals took hours, and I was always exhausted when they were over. I realized then that I could either use my time to feed myself or have an attendant feed me, allowing me to spend the time saved to go to school. I went to school.

## A MOVE TO EMPOWERMENT

Groups of disabled people finally began to question the role that society had assigned them. One of the first groups to assert themselves was the blind population. In the 1930s the California Federation of the Blind was founded in Berkeley, followed by the foundation of the National Federation of the Blind. These organizations were run and controlled by blind persons and reflected their aspirations to create a better life for themselves in society.

In the 1950s and 1960s a new breed of organizations was formed around the country; it was a departure from the traditional organization run *by* able-bodied persons *for* people with disabilities. People with disabilities found that by forming their own associations and directing and staffing them themselves, they were assured of consumer control. Some associations, like Disabled in Action, still exist; others, changed their names. For example, the New England Spinal Cord Injury Foundation became the National Spinal Cord Injury Association. Some groups, like the American Coalition of Citizens with Disabilities, went under. These and similar groups have played an important role in the disability rights movement, because they looked at issues from the disabled person's point of view.

Before the civil rights movement, independent living meant being able to manage daily living tasks either alone or with somebody's help. In contrast, the independent living movement that emerged in the 1970s was a political concept with a clearly defined social model for the integration of people with disabilities into their communities. It embraced the normalization philosophy outlined by Wolf Wolfensberger (1977) and the civil rights and equal opportunity doctrines of the 1960s.

## INDEPENDENT LIVING AND CIVIL RIGHTS

One evening in 1976, televised news dramatically introduced millions of Americans to the new civil rights movement of people with disabilities. Simultaneous demonstrations in San Francisco, New York, and Washington, D.C. demanded the signing of regulations to implement Section 504 of the Rehabilitation Act of 1973, the civil rights act for people with disabilities, which states that "no otherwise qualified handicapped individual . . . shall, solely by reason of his handicap, be excluded from the participation in, be denied the benefits of, or be subjected to discrimination under any program or activity receiving federal financial assistance." (Federal Register, Wednesday, May 4, 1977).

People with many kinds of disabilities—deafness, blindness, cerebral palsy, spinal cord injuries, mental retardation, mental illness, and multiple disabilities—moved into federal buildings across the country and refused to leave. In the San Francisco Bay Area, the demonstrators organized a 28-day takeover of the federal building, the longest takeover in the history of the United States. It was no coincidence that the most effective sit-in occurred in the birthplace of the independent living movement. Here, the coalition of people with varied disabilities was a working reality, not a theory. The guiding principle was that once civil rights were guaranteed, independent living could be achieved. These demonstrations culminated in one of the more significant victories in the history of the civil rights movement, the signing of regulations (PL95–602) to implement Section 504 of the Rehabilitation Act of 1973 by Secretary of Health, Education and Welfare Joseph Califano. He summarized the importance of this law as follows:

Section 504 . . . represents the first Federal Civil Rights law protecting the rights of handicapped persons [sic] and reflects a national commitment to end discrimination on the basis of handicap. The language of Section 504 is almost identical to the comparable non-discrimination provisions of Title VI of the Civil Rights Act of 1964 and Title IX of the Education Amendment of 1972 (applying to racial discrimination and to discrimination in education on the basis of sex). It establishes a mandate to end discrimination and to bring handicapped persons [sic] into the mainstream of American life. The Secretary intends vigorously to implement and enforce the mandate. (Federal Register, May 4, 1977)

In 1975 the fundamental educational rights of children with disabilities was guaranteed through Public Law 94–142, the Education for All Handicapped Children Act of 1975. This law addressed two fundamental problems: (1) the isolation caused by segregating disabled children in the school system and (2) the denial of education altogether because of lack of programs. Prior to this act, 4.5 million disabled children were excluded from the school system. Now parents, the child, and society could have the same expectations for a good education that have been enjoyed by nondisabled children. Integrating disabled children into the school systems would change attitudes among disabled and nondisabled young people in a socialization process benefiting the entire student body. The law also recognized parents as primary advocates for their children and provided them with rights to influence the now-required individual education plans.

In 1974 the disabled community strengthened its clout by funding the American Coalition of Citizens with Disabilities (ACCD), which provided a unified voice for about 150 different disability organizations.

## THE CENTER FOR INDEPENDENT LIVING

The concept of independent living, born in the early 1970s in California, evolved from the Disabled Student Program at the University of California, Berkeley, which in turn led to the founding of the Center for Independent Living (CIL) in 1972.

The Disabled Students Program (DSP) had its roots in the Cowell Residence Program at the university; it allowed severely disabled students to attend classes and reside on campus. The DSP was a self-help advocacy program run by people with severe disabilities. This group of disabled individuals pushed for, and obtained, curb cuts and ramps on and around the campus. But a grave problem remained: Once graduated, where could these disabled people live—or work? How could they effectively move about their communities?

The students in the DSP conceived the idea of "independent living," which to them meant active participation in society—working, having a home, raising a family, and generally sharing in the joys and responsibilities of community life. Independent living meant freedom from isolation and institutionalization; it meant the ability to choose where to live, how to live, and how to carry out the activities of daily living that most able-bodied people take for granted. It meant taking the responsibility for political action and charting a new way of life. The Disabled Student Program at Berkeley was the testing ground for the independent living philosophy.

The first Center for Independent Living (CIL) was founded by these same people. It was initially funded by a grant from the University of California, Berkeley. The center followed the student model in being staffed by people with a wide range of severe disabilities. It supplied services such as attendant referral, financial aid counseling, architectural barrier removal, and individual advocacy for grievances with city, county, and state governments. In time the CIL expanded to include more peer counseling, readers for the blind, a list of interpreters for the deaf, peer support for the mentally ill, wheelchair repair, van modification, and teaching independent living to mentally retarded people. Eventually, CIL taught independent living to the world.

Because of its funding requirements, CIL was of necessity devoted to service; because of its commitment to change, it was also devoted to advocacy. The center did not charge for its services. Employees of the center earned only as much as was allowed within the limitations of various disability benefits. They could not afford to take full-time pay for full-time work because it would not make up for the loss of medical benefits.

By 1974 CIL needed more space. It moved into a rickety old building with an elevator from which the fire department periodically had to rescue people. People from around the Bay Area needed independent living services. The budget for CIL increased with its funding, which included a federal grant from the Rehabilitation Service Administration after the Rehabilitation Act of 1973 was signed.

CIL gained prominence through advocacy. Its staff helped shape public policy by testifying before state legislatures and agencies. The disabled consumers of CIL addressed issues like attendant services, medical benefits, and work incentives. They lobbied for funding of technological assistance devices through Medi-Cal and medicare. Every available tool of advocacy was used, including the courts.

CIL moved to its present address in 1975. By this time, the center had attracted international attention. People came to Berkeley from around the world to see independent living in action. Some people with severe disabilities even hitchhiked.

One of CIL's strategies was to use each job at the center to train people who were hitherto excluded from the labor market as being too disabled and lacking in experience. With training, people from CIL moved on to establish other centers and to find jobs in governmental policy-making positions. My own career is an example. When I left CIL as director in 1975, I was appointed director of the California Department of Rehabilitation and became the first severely disabled person ever to be named to this position.

The development of the Center for Independent Living was a milestone in the short history of the civil rights movement for people with disabilities. Disabled people around the world embraced the concepts of self-determination, independent living, and integration into society and implemented them through cost-effective self-help.

The 1978 amendments to the 1973 rehabilitation act provided some federal funds for the independent living movement and helped to establish independent living centers in all 50 states. By the end of 1987, there were over 400 independent living centers nationwide and 100 outside the United States.

## INDEPENDENT LIVING PROGRAMS

Independent living programs, which developed throughout this country, are modeled after key features of the CIL in Berkeley. While they differ in organizational form and range of services, they tend to share the following basic characteristics:

1. *Self-determination.* Disabled people must be involved in the development of services to meet their needs. CILs promote a high level of consumer involvement. Most of the programs are managed or directed by people with severe disabilities and many of the staff members are disabled. By federal mandate, a majority of the board members of any CIL must be persons with disabilities.

2. *Self-image and public education.* Independent living centers use peer support, self-help concepts, and disabled role models to make the disabled individual aware of his or her potential for a life of greater participation and active involvement. To counteract the low expectations the community holds for people with disabilities, independent living programs and centers

begin with high expectations and a "can do" attitude. At the same time, independent living programs educate the public about the potential of people with even the most severe disabilities to live independent, productive lives with dignity and respect.

3. *Advocacy.* The methods of CILs focus on advocacy, which fights discrimination based on disability and promotes independent living, equality of opportunity, and full participation of people with disabilities. For example, the independent living movement agitated for a national policy to make transportation systems accessible. It got a commitment from the Carter administration that in the future all buses funded with federal money would be fully accessible. Unfortunately, the commitment wasn't honored by the Reagan administration. The CIL advocacy is aimed at adapting existing services rather than creating a new service for the exclusive use of disabled people. "Separate but equal" is not acceptable.

4. *Service to all.* CILs serve people with all disabilities, of all ages, as well as their families. They provide services according to need rather than the category of disability. A crucial difference between the independent living philosophy and the rehabilitation model is that services are not limited to short-term need. They are provided for the lifetime of disabled individuals.

Independent living programs offer choice and control, thus fostering the empowerment of persons with disabilities. CIL services assist people to move from segregated classrooms, isolating hospitals, convalescent homes, and sheltered workshops to freedom. The medical system has done a superb job in saving lives, but it is up to people with disabilities and their families and friends to advocate policies which will further consumer control, empowerment, and cost-effective alternatives to dependency.

In California, independent living has proven cost-effective. People who are highly motivated and involved in their communities cannot afford to be sick. People who are dependent and isolated will often choose sickness and subsequent hospitalization because it frees them from society's demands. In 1980 a California study by Berkeley Planning Associates reported that people served by independent living programs spent considerably less money for medical services than a similar group of people not served by these programs.

## DISINCENTIVES

In the past few decades, people with disabilities who have moved from a welfare mentality to one that has seen them become contributing, productive members of society, have shown that motivation can overcome perceived limitations. The legal disincentives that prevent people from working and living in the community must be changed.

The economic system of supporting people with severe disabilities can be described as an all-or-nothing system: earn a little and lose a lot. The most crucial loss can be that of medical benefits. The fear of losing medicare/medicaid coverage because of prohibitions on employment income often prevents people with disabilities from taking the step toward independence. Independence should be rewarded through a system of incentives for those who wish to leave the welfare rolls. An example of progressive legislation is Section 1619 of the Social Security Act, which allows people who go to work to keep a greater part of their new income, while retaining basic medical benefits and other vital assistance.

It is less expensive to keep people in their own homes than to institutionalize them. The insurance industry offers few programs that allow people to remain in their homes supported by personal assistance, thus forcing them into more restrictive and more expensive environments like nursing homes. Changes in current policies in order to promote independence would be beneficial to insurance companies and consumers alike.

It is ironic in an age when technological aids can compensate for the loss of arms, legs, sight, and hearing and actually provide opportunities for jobs, that few government or private agencies will purchase this equipment or pay for its maintenance. For example, a computer can make a quadriplegic productive in his or her personal and work life, yet few agencies will purchase one. A major hurdle is that all equipment must be prescribed by a physician and be justified as "medically necessary." Thus, communication devices for deaf people or for people unable to speak are denied as not medically necessary.

We must redesign and rethink business practices in the area of disability. We must reevaluate the role of professionals. The peer counseling model should get wider currency since people with disabilities are demonstrably more effective in counseling each other. In addition, it is less expensive for a person with a disability to counsel another with a disability than for a professional to do so.

## COALITIONS AND THE 1980s

In the 1980s disabled people have joined together to form new coalitions for advocacy based on the independent living philosophy. One such organization is the Disability Rights Education and Defense Fund (DREDF), a spin-off of the Center for Independent Living in Berkeley, which uses the legal system and attorneys with disabilities to fight for the rights of disabled people.

The more than 350 federal, state, and privately funded independent living programs in the United States have formed the National Council of Independent Living (NCIL). This organization is successfully working to increase funding and commitment to independent living.

At the policy level, the National Council on the Handicapped (NCH) was created to present policy alternatives to the federal government. Its mandate is to help bring order to the existing "nonsystem" of transportation, housing, welfare, rehabilitation, and employment programs for disabled people.

Disabled Peoples' International (DPI) represents a coalition of disabled people and their organizations in 70 countries. This group aggressively promotes basic human rights and the independent living philosophy. Its goal is to unite the more than 600 million people with disabilities worldwide. DPI has advocated for disability issues before such international organizations as the United Nations, the World Health Organization, and the International Labor Organization. It has moved into developing countries to train disabled people as leaders in the politics of disability and the art of independent living.

In 1983, my interest in the development of the movement led me and other civil rights leaders to form the World Institute on Disability (WID), a public policy and advocacy organization. WID's agenda for this decade is to contribute toward a comprehensive public-policy approach in the disability field. The institute has focused on four programmatic areas.

**1.** *Personal Assistance Services.* The availability of attendant services is a requirement for many severely disabled people. This service often makes the critical difference in deciding whether an individual can work or live in the community without the need for placement in a nursing home or other restrictive environment.

**2.** *Public Education.* The term disability can evoke fear and pity. The institute is working with the media and is producing materials that will enable the public to learn about attendant services, employment, independent living, health care, and aging. People with disabilities want to be seen first as people, not as the personification of disabilities.

**3.** *Service System Consultation.* The institute provides consultation to numerous organizations and agencies on a variety of issues, such as accesssible public transportation, aging and disability, housing, legislation, and employment.

**4.** *International Development of the Independent Living Movement.* WID is actively involved in promoting relations among disabled communities in different countries. Opportunities to visit other countries and to host visitors from other countries have led to an exchange of knowledge about disability and related issues. WID has developed an international exchange program to assist people from other countries who are interested in learning about program innovations in the United States.

## THE NEXT STEP: COMMENTS ON CURRENT ISSUES AND PROBLEMS

There are serious concerns about the long-term viability of independent living programs. Will they become bureaucracy-bound and lose sight of their origins in the grass roots? Some programs are losing their political will and dropping advocacy for fear of economic reprisal. Funding sources must be diversified to assure survival of independent living programs. No one funding source must control a center whether the source be governments, corporations, or foundations. As part of a diversified funding strategy, the centers should bolster their economic self-sufficiency by developing businesses such as wheelchair shops and orthopedic and medical supply stores.

Filling job positions within the disability movement with disabled people, who had learned and could translate their experiences into action for positive change, was the most important issue in the disability movement some 20 years ago. Nondisabled people were not excluded, but priority was given to hiring people with severe disabilities, people who had been systematically excluded from training in management skills and organizational development. At present, many independent living programs around this country are directed by nondisabled people. The excuse "We could not find a qualified disabled applicant" rings hollow when there are hundreds of gifted and competent people with disabilities who are ready now to assume leadership positions.

Participants in the movement must learn to confront problems without undermining people's dignity. Conflict is acceptable, even necessary, but it needs to take place in a fair and supportive context. There is room for all in this movement, including people with mental disabilities, people with disabilities who live in rural areas and Third World countries, elderly people, and children.

Some independent living programs have not reached out to people with other than physical disabilities. The talents emerging from self-help groups of mentally and developmentally disabled people are only beginning to be used. The independent living movement is just emerging in rural and minority communities. More aggressive action is needed in order to bring people from all racial groups into leadership positions in the movement recruitment practices and in locating new independent living programs. It must be recognized that increasing numbers of disabled people are old. We must design our services to empower them and to prevent their automatic institutionalization. Finally, we must give more attention to children with disabilities. Role models and peer counseling are critical to teaching parents and their children how to advocate for the children's future and how to give them a vision of what is possible.

## SUMMARY

For society to warehouse people with disabilities by segregating and isolating them is no longer intellectually or morally acceptable. Society loses too much by ignoring the talent, creativity, and potential value of this large segment of the population. People with disabilities have shown through the success of independent living what contributions they can make to society. However, a wide range of services is necessary to enable a person with a severe disability to live independently. Each severely disabled person has unique service needs and requirements. Individual choice is vitally important.

The removal of barriers to normal living—whether in housing, transportation, employment, education, or in attitudes—is a prerequisite to self-determination. Assumptions that disability necessarily means dependence, nonproductivity, and powerlessness have to be rooted out. The independent living movement has fought and will continue to fight these attitudes by rewriting old laws and by providing incentives for people with disabilities to work, to live outside institutions, to go to regular schools, to move about freely, and to take advantage of technological advances. If accomplished, such changes would lead to a higher quality of life for everyone.

In summary, the independent living model offers higher societal benefits, creates more opportunities for dignified human interaction, and is more cost-effective than the traditional medical and institutional models. Independent living programs and their philosophy have proved a success.

The challenge is to actualize these models effectively. Every person with a disability must have access to an independent life.

The problems of how to live with a disability is a problem for all members of society. Our culture needs to evolve to the point where even the least empowered are able to develop the skills necessary to live in the community. At such a time, society will reach its highest potential.

## REFERENCES

*Federal Register.* (Wed., May 4, 1977). Department of Health, Education and Welfare, non-discrimination on basis of handicap (part 4) *42*(86), 22676.

Fuchs, V. R. (1982). What is the cost-benefit analysis? In T. L. Beauchamp & L. Alters (Eds.), *Contemporary Issues in Bioethics.* Belmont, Calif.: Wadsworth Publishing, 476–477.

Johnston, V. M. & Keith, R. A. (1984). Cost-benefits of medical rehabilitation: Review and critique. *Archives of Physical and Medical Rehabilitation, 64*(4), 147–154.

National Center for Health Statistics, (1986). Current estimates from the National Health Interview Survey, United States, 1985. By A.J. Moss & V.L. Parsons. *Vital and Health Statistics,* ser. 10, no. 160. DHHS pub. no. (PHS) 86–1588.

Roeher, G. A. (1961). Significance of public attitudes in the rehabilitation of the disabled. *Rehabilitation Literature, 22*(3), 66–72.

Stoddard, S. (1980). Evaluation report on the state's independent living centers. Berkeley Planning Associates, Berkeley, Calif. (Funded by AB 204; final report submitted to the California State Department of Rehabilitation)

Thomas, S. & Wolfensberger, W. (1982). The importance of social imagery in interpreting societally devalued people to the public. *Rehabilitation Literature, 53,* 356–358.

Verville, R. E. (1979). Federal legislature history of independent living programs. *Archives of Physical Medicine and Rehabilitation, 60*(10), 447–451.

Wolfensberger, W. (1977). The normalization of people, and some major implications of architectural-environment design. In M. J. Bednar (Ed.), *Barrier-Free Environments.* Stroudsburg, Pa.: Dowden, Hutchinson and Ross, 135–169.

# 15
# Discrimination and Disability

## Albert R. Jonsen

The field of ethics in medicine is largely concerned with the issues of death and dying. Disability has been discussed solely in relation to the birth of a disabled child, questions that arise in the daily life of disabled persons and in the care of the disabled have not been extensively treated in the field of medical ethics. Thus, I cannot fall back upon a volume of scholarly work that compares to that of the death and dying issue.

Recently, the New York Times reviewed a book entitled *Lisa H., The True Story of An Extraordinary and Courageous Woman* written by Richard Severo (1986), a journalist. It is the true story of a young woman who suffered from neurofibromatosis, the disease said to afflict Joseph Merrick, the "Elephant Man." This young woman had a severe form of this disease that terribly disfigured her face. The story relates her preparation for a difficult but only palliative surgical procedure. The reviewer writes that the story is vividly and beautifully told, bringing out the extraordinary courage of this young woman. "We must first watch Lisa endure her tormenters, the 'turkeys' as she calls them. Fortunately, for every bus driver who whispers 'You're the ugliest thing I've ever seen in my life,' there is an innocent child who unknowingly supports her. Told by his second grade teacher to make believe Lisa is lost and the police need a description, a classmate describes her as having 'brown hair, hazel eyes and a certain height and that's all'" (Schrier, 1985, 11).

The story of Lisa H. illustrates the discrimination encountered by the disabled. The etymologies, or original meanings of words, often shed great light on current meanings, which have drifted away from them. For example, we now think of *discrimination* primarily as an illegal act that bars the access of certain classes of people to certain benefits in society. However, in its etymology, discrimination basically means making a judgment. It comes from the Latin word *discrimen* and a Greek word that means separating, distinguishing. Discrimination means making a difference, showing a difference, judging that one thing is different from something else.

The real issue is not whether there is discrimination against a group or against a person but whether that discrimination is morally right or wrong. We speak of discrimination as being "invidious" discrimination, and we mean that the separation, the differential treatment, is not morally justified. "Invidious" comes from the Latin word *invidia*, which means envy—thus, envious discrimination. But, envy of what? Why would one envy a person one is discriminating against? The answer seems to be jealousy and the jealous individual, who wants to take from the disabled their dignity and their rights. This is what makes discrimination evil. It arises from a jealousy that someone in your presence is claiming to be a person.

Some forms of discrimination, however, are quite ethical. In education, for example, it is important to discriminate in favor of children who are slower learners. One should separate them in ways that will assist them to catch up and to achieve adequate levels of performance. We also usually consider it fair to discriminate in favor of brighter children, though this is a more complex issue. We discriminate in favor of the sick when we give them care. On the face of it, we see certain individuals, certain situations in which special efforts are clearly justified. We seek to identify those for whom special treatment and special efforts will provide equal opportunity with others. This is ethical discrimination. We might argue about what constitutes equality of opportunity in each particular case, but it will always involve treating some people differently in order to bring about development, progress, and improvement, just as invidious discrimination seeks to demean others and to uphold rigid separation. Slavery and caste systems are examples of the latter.

Thus, ethical or moral discrimination aims for equality of opportunity and seeks positive development. This is the discrimination that is built into the definition of justice. The oldest definition of justice in our culture goes back to the Greeks: Give to each one according to his due. This is discrimination: it assigns what is owed to this person and what is owed to that person. The general definition of justice is a discrimination between persons, each of whom is owed certain benefits and must bear certain burdens in life.

If we look to the issue of justice and the disabled, we see certain notable developments in recent years. In the legal arena, we see a series of legislative enactments and judicial decisions that have attempted to provide equality of opportunity for the disabled in significant ways. The legislation of 1973 and 1976 is an effort to establish equality of opportunity and to bring about the changes that are due those who suffer from special disadvantages within the society. It resulted from extraordinary efforts by individuals and groups to deal with injustice relative to the handicapped. These efforts had crystallized into the Independent Living Movement, which has worked with great energy to dismantle the barriers to equality of opportunity. Yet despite this legislative progress, despite the concern of many individuals, there remains enormous discrimination against the disabled in this society. The reason for this lies in the cultural and social perceptions of disability, which value and devalue people in certain ways. We have failed to overcome these in any major way.

It seems this enormous discrimination is an inevitable part of human existence, a natural way in which human beings relate to one another in every society that we know. Many societies, particularly as they become more complex, create despised classes. One has to look carefully at the sociology of a society to see how this comes about. The etymology of the word *despise* is of interest. It derives from the Latin *despicere*, which means to look down upon. The immediate implication is that certain persons are held in contempt and are treated thus, as in caste system in India. Another meaning of despised is more relevant to the disabled in the United States. "Looking down upon" means looking from on high, and when one looks from on high, one can also overlook. A despised class may not necessarily be a class held in hatred or contempt but a class of human beings that are overlooked. Society looks away from them. It looks away from them in embarrassment but does not look at them, does not see them for what they are and, more importantly, what they can be in their contribution to society. In this sense, the disabled are a despised class.

The following discussion will not make precise distinctions between particular disabilities. The general analysis can be adapted to almost all of the major disabilities that individuals may suffer. Six features of the social perception of disabled people begin to account for the creation of this despised class and for the great difficulty that both legislative and individual initiatives have had in dismantling the barriers to equality.

The first feature is that people who are disabled from birth—who suffer from genetic or congenital disabilities—are identified with their disease, that is, the condition did not come upon them late in life. If a person is found at the age of 55 to be atherosclerotic and to have clogged coronary arteries, the disease is something that happens to that person. The congenital or genetic defect didn't happen, it was there, and the person is identified with that defect or disease. This spurious identity is the first source of invidious discrimination: the disease itself is evil, and therefore, in some sense, the one identified with it is evil. These are attitudes that underlie social perceptions. We would, of course, never admit this. In fact, we disguise these attitudes more thoroughly by attempting to get rid of any language that would even imply them. The use of sanitizing language oftentimes hides deeply invidious forms of discrimination. Sometimes we undermine ourselves by getting rid of the offensive language; it may be a good thing to have this language around because it makes clear where the evil lies.

The second problem in the perception of the disabled as a despised class is that disability is very often thought to be associated with mental incapacity. Sometimes disability is indeed associated with mental incapacity, as when individuals suffer from diseases that affect their mental abilities in a profound way. Certain disorders affect metabolism so that the brain is profoundly damaged. Or it can be damaged by anoxia at birth. In either case, there has been genuine impairment of mental capacity. In other disorders there is no actual impairment of mental capacity but only a perceived one. The strange cultural association of physical limitation with mental limitation is probably most striking in the case of people suffering from cerebral palsy, in which individuals who are physically limited (especially if they have a speech articulation defect) are assumed to be psychologically or mentally limited. In most cases this is simply false. The assumption is terribly harmful in a society that prizes intelligence. It can lead us to question whether we value intelligence in the sense of wisdom or only in the sense of "smarts". Having smarts is much more about being able to move quickly than it is having any true understanding of the world. If one can't move quickly, then one can't be smart.

Third, disabilities are clearly associated with dependencies, and this society values independence and productivity. Again the question is, what sort of independence do we prize? We all are aware that every human being is dependent in profound ways; it is our denial of those dependencies that oftentimes gets us into trouble. Yet, for good reason, the disabled have more visible dependencies than other persons. In addition, they do not appear to be productive in the sense of the competitive productivity that society prizes and are thus dismissed as potential earners and taxpayers. From the ethical viewpoint, severe criticism should be brought to bear against our conceptions of productivity and of independence. The contribution of disabled individuals must be measured not only in terms of finances but also in terms of their contribution to those with whom they live and work. Assuring their equal opportunity is a contribution to society's sense of itself and hope for the future.

Fourth, disabled persons present themselves as different. Indeed they are different, because they need different kinds of assistance. Yet, in our society uniformity is a highly prized cultural phenomenon. Tragically for disabled people, it is a uniformity that stresses anonymity: no one should stand out as being different, and therefore one need not pay attention to those who do. This is a common criticism of urban life, nevertheless, anonymity is prized, and one must show no signs of being distinct or different. The disabled, however, always show those signs. They cannot help but carry those signs with them.

Fifth, disability evokes—or should evoke—compassion, in its literal, etymological sense, that is, feeling *for*, feeling *with*. It is the ability to say, "I can understand in some remote way how you feel." Those who work with the disabled must be people of compassion. One can never identify with disabled persons, one does not live with the disability as they do. But one can empathize. Those who work with the disabled in the most skillful way are people who have achieved a high level of that kind of compassion. One can say, "I can now imagine what it is like to encounter these barriers to opportunity, to physical movement, and so forth." Yet, in our society, compassion is an embarrassment. Its connotation of pity overwhelms its meaning as *feeling with*; it becomes a despised virtue. A *New Yorker* cartoon shows the office of a company president with a large window in the back and smokestacks outside the window. The great tycoon is staggering; he is holding his hands up to his head; two young men are propping him up on either side as he says, "I was suddenly overcome by a wave of compassion for the poor."

The final social perception of the disabled person is bound to give rise to some disagreement. It is a complex and distressing topic yet unavoidable, for it has an impact on the disabled and will in the future have even more of an impact. It is the perception of the disabled as disposable. The best example of disposability in our society is Kleenex: something that can be thrown away without loss. One reason the disabled are perceived as disposable is society's position on abortion. It would be a different matter if abortion were entirely a question of a choice arising from the distress of a woman carrying a child she did not wish to carry and whose identity and characteristics she did not know at all. An abortion where the identity and characteristics of the child are hidden behind a veil is to me morally more acceptable than abortion where the characteristics of the child are known. The abortion based on information derived in prenatal diagnosis is most troubling. I am torn about this issue. On the one hand, parents suffer great distress when they suspect their fetus is at risk, and on the other, the destruction in utero of the child with disability may be the basis for a general feeling of disposability toward the disabled, which is most troubling. Almost everyone feels strongly about this highly debatable question; I could not honestly refrain from including it.

These are some of the reasons that the disabled are a despised class in this society. None of the reasons are based on empirical facts but on impressions from my limited experience. These social perceptions of disability—disability as disposability, disability as evoking embarrassing compassion, disability as being strikingly different, disability as dependence, disability as mental incapacity, and disability as an identity with disease—are the hindrances to justice for the disabled. They are a hindrance to justice because they perpetuate the institutions that continue to distribute benefits and burdens in unequal ways to the citizens of society. But even more, they are a hindrance to the most fundamental of all human goods—self-esteem. Without self-esteem, none of the other goods and benefits of life can have meaning. To be a member of a despised class is to suffer continual assaults on one's self-esteem. Indeed, efforts have been made toward normalization, from activities such as the Special Olympics to the legal removal of physical and occupational barriers. Still, it seems that in the absence of institutions that promote self-esteem, those who suffer from this invidious misperception and belong to the despised class need to be persons of enormous courage. This may sound like a platitude; it has been said many times. It is spoken of in the book "The Story of a Courageous Woman." Courage is a person's self-esteem under pressure from social institutions that do not appreciate that person. If we apply this to the disabled, we can say that the fabric of their self-esteem is threatened on all sides in a society that does not value their difference, their independence, and their dignity. They can function only with courage.

Philosophers must oftentimes leave people without answers. This author has no answers; the reader may already have answers that go far beyond the questions raised in this chapter. It is hoped that this discussion will stimulate some thoughts on how certain social perceptions and misperceptions may be related to society's inability to come to grips with the presence of disabled people in its midst.

## REFERENCES

Schrier, E. W. (1985). Heroic surgery, heroic patient. *New York Times Book Review,* 10 February, 11.

Severo, R. (1986). *Lisa H.: The true story of an extraordinary and courageous woman.* New York: Penguin Books.

# Index

ABLEDATA, disability information retrieval service, 172
abortion, prenatal diagnosis and, 250
adaptation to disability, 56, 170
advocacy, 239, 240–241
American Coalition of Citizens with Disabilities (defunct), 235
antilocalization theories of brain function, 58
anxiety, 54
appearance: effect of changes in, 14. *See also* facial disfigurement
Arthur point scale, 67
Assistive equipment: advocacy's impact on, 172; and assessment, 178–179; background, 171–173; and competitive sports for disabled, 172; consumer preferences, 179; developmental assistance offered by, 173–174; environmental changes impact on use of, 172; functions of, 170–171, 175–176, 179; hindrances to provision of, 172–173; increased use of, 171; and independent living, 172–173; improvement via technology, 171–172; legislative changes related to, 172; selection of, 174
attention, defined, 67
attention deficits: assessment strategies, 69–70; causes, 68, 69; versus cognitive deficits, 67

Baby Doe controversy, 222
Barnum, P. T., 152
Becker muscular dystrophy, 109
behavioral medicine, xi
Bender-Gestalt Test, in neuropsychological assessment, 59

Benton Visual Retention Test, with homonymous hemianopsia, 64
body "armature," 10
body: attitudes toward, 6–7, 8; boundaries, 12; conversion reactions, 9; compared with others' bodies, 13–14; connections with mind, 5–7, 8; depersonalization of, 14–15; as embodiment of self, 7, 16; expression and communication, medium of, 7, 24–25; fears about destruction, mutilation, or malfunction, 10; fears about disempowerment and loss of competency, 22–24; hyper- or hypo-awareness of, 15–16; idiosyncratic ideas about 9–10; integrative view of 7; limitations of, 21–22; libidinization of parts of, 11; medical model, view of, 6–7; misrepresentation of size and shape of, 15; moral feelings about, 14; optimalization of performance of, 25; plasticity of, 8; projections of conflicts about onto others, 11; psychodynamic attitudes toward, 6–17; psychological experience of, 9–12; in psychotherapy, 8; psychosomatic ailments, 9; self-integration, symbol of, 14–17; self-mutilation, desire for, 10–11; self-worth, symbol of, 13–14; social interaction, problems related to, 12–13; as thing-like, 21; and unconscious conflict, 7, 21
"body ego," 20
body image, 15, 19, 22; in cerebral palsy and spina bifida, 89; in dreams and fantasies, 11–12; in illness and disability, 16, 22–26; as source of psychological conflict, 20–21

253

body schema, 6, 7, 15, 19, 20, 21
Bouvia, Elizabeth, 227
brain lesions, and personality disorders in children with cerebral palsy, 90–91
*Brown v. Topeka Board of Education*, 224

Califano, Joseph, 235
California Federation of the Blind, 234
capital disability rating scale, 140
Center for Independent Living: activities, 237; history, 236–238; job-training, 238. *See also* independent living movement
cerebral palsy, incidence, 81
cerebral palsy and spina bifida: biologic factors in personality development, 88–91; early development, 88; influence of psychosocial factors, 91–95; prognosis, factors in, 87
compassion, etymology and meaning, 249
control/mastery and disability, 129–130, 177–178
conversion reactions, 9
Cowell Residence Program, University of California, Berkeley, 236
cystic myelomeningocele sac, impact of surgery on for personality development, 91

decision-making for disabled child, family's role, 94
denial, 55; in neurological patients, 55; in multiple sclerosis patients, 138; in spinal cord injured patients, 121
dependency, 54
depersonalization, 14–15
depression, 53–55. *See also* facial disfigurement; multiple sclerosis; spinal chord injury
Digit Span (subtest of Wechsler Adult Intelligence Scale), and attention deficits, 70, 71, 72
disability: adaptation to, 45–46, 50; defined, 176; discrimination against persons with, 232, 245; family reactions to, 46, 47, 50, 51; increasing prevalence of, ix, 232; misimpressions about, x, 44–45, 232–233; phenomenological overview, 5–29; and self-esteem, 250; social and legislative changes regarding, x; as stressful life event, 45; treatment and rehabilitation of, 50; versus handicap or impairment, 223
Disability Rights Education and Defense Fund (DREDF), 241
disabled as a despised class, 247–250. *See also* stigma
disabled child, reactions to, 30
Disabled in Action, 234
Disabled Peoples' International, 241
disabled rights movement, and empowerment, 234, 235. *See also* independent living movement
Disabled Students Program, University of California, Berkeley, 236, 237
discrimination: and disability, 245; etymology and meaning, 246; in employment, 225; institutionalization of, 232; invidious versus ethical, 246
distractibility, freedom from, 73. *See also* attention deficits
Doe, Baby, controversy, 222
Duchenne muscular dystrophy: cerebral origin of, 108; clinical course and symptoms of, 108–110; diagnosis of, 110–111; family reactions to, 111, 113; genetic counseling for, 109; group psychotherapy and, 115; incidence and prevalence of, 82, 109; parents' reactions to, 113; physical therapy for, 111, 113; special education, 113; sports, 113; surgery for, 114; therapeutic management of patient, 111–116; therapeutic management of parents after child's death, 116
dystrophin, 109

# INDEX

educational model, in rehabilitation, 130–131
Education for All Handicapped Children Act (PL 94–142), 224, 236
Elephant Man, 245
Eliot, Charles William, 147
embodied self, 7, 19–21
empowerment, of disabled persons, 234
ethical issues in disability: autonomy, 227; charity versus scarcity of resources, 227; discrimination in employment, 225; "duties of charity," 227; equal opportunity, 225; equity versus efficiency, 226; liberty versus equality, 224; medical care of disabled, 222; moral responsibility, 223, 228; social policy, 224
euphoria, in multiple sclerosis, 137

face, psychological and social importance of, 148
facial disfigurement: adjustment to, 157; attitudes of nondisfigured toward, 150–153; attitudes of professionals toward, 151; cosmetics and prosthetic appliances for, 160; employers' attitudes toward, 156, 158, 168; family reactions to, 154–155; family therapy for, 159; and identity, readjusting sense of, 153, 155; impact on others, 148, 149, 152; impact on self-concept, 154, 155; impact on social interaction, 150; impact on social self, 154–155; innate versus acquired fear of, 152; and litigation, 157; and long-term goals, 157; managing apathy and depression about, 153, 156; medical model, 156; peer groups, 160; pharmacological treatment of patients with, 159; psychotherapy for, 159; rating scale for severity of, 150; reactions of school-age peers toward, 155; and sexual functioning, 158; sociological perspectives on, 149–151; as a stigma, 150–152;
and stress, 151; surgical interventions for, 159; therapeutic interventions for, 155–158; family influence on child's emotional status and personality development, 91; role in decision-making for disabled child, 94
forensic psychiatrist, functions and role, 200–201, 203–205
forensic psychiatry, and disability: answering questions of causation, 206, 207; challenges of testimony, 211; challenges to professional integrity, 213; deciding who can work and who cannot, 214; determining compensation, 208–209; diagnosis of disorders, 206–207, 210–211; and DSM III-R, 210–211; forensic epistemology, 207; litigation, 201, 216–217; psychiatric exam, 209–210; post-litigation stage, 217–218; quantifying mental distress and disability, 213–214; and rehabilitation, 216; social values versus medical factors, 214–215

Gotham Proverbs Test, 70
Guinea Pig Club, 160

Halstead-Reitan Neuropsychological Battery, 59, 60
handicap, defined, 176; versus disability or impairment, 223
health psychology, xi
hemangioma, 147
Hiroshima survivors, reactions to facial disfigurement, 153
homonymous hemianopsia, 64
Human Sexuality Program, University of California, San Francisco, 184
hydrocephalus, impact of shunt insertion for, 91
hyperverbal behavior, due to organic deficits in children with cerebral palsy, 89
hypoxic encephalopathy, 91

impairment, defined, 176; versus disability or handicap, 223
Impaired child, treatment and rehabilitation of: accepting limitations, 35; collaboration between parents and professionals, 31, 32, 36; decision-making, 32, 35, 37; legal guidelines, 34; multidisciplinary team and, 31; uncertainty of, 35
Independent Living Center programs, functions and goals, 238–239
independent living movement: disincentives to, 239–240; future steps, 242; philosophy, 235; promotion of, 231, 235–236; versus rehabilitation model, 239
information retrieval services: ABLEDATA, NARIC, 172
institutionalization, of disabled, 232
intervention in rehabilitation, medical versus social learning model, 130–131

Jobst mask, 148
justice: and compassion, 229; distributive, 224; ideal versus conservative, 226; poverty of, 228

Knox Cube Test, 67, 68, 69, 70

learned helplessness, in spinal-cord–injured patients, 129–130
legislation and disability, 246–247
Leiter International Performance Scale, 65
Luria-Nebraska Neuropsychological Battery, 73

Masters and Johnson's Institute, and treatment of sexual dysfunction in disabled persons, 190
medical model: and disability, 231–232; emergence in 1940s, 233; with facially disfigured patients, 156; in spinal cord injury rehabilitation, 130; narrow focus, 6; versus educational or social learning model, 130–131
Merrick, Joseph, the "Elephant Man," 245
Minnesota Multiphasic Personality Inventory (MMPI), 125, 127–128, 139; fallacy of average MMPI results, 128
motivation, lack of, 54
motoric slowing, 62
multiple sclerosis: adjustment, psychological factors in, 140; cognitive deficits and, 139–141; denial in, 138, 140, 143; diagnosis, 138; emotional reactions to, 139, 141, 143; environmental manipulations and, 145; euphoria in, 137, 139; hysteria in, 139; integration of patient into home and community, 145; "MS personality," 137–139; onset, 137, 138; pharmacological treatment of, 144; psychotherapy, individual and group, 145; prevalence, 82; and sexual dysfunction, 139; stress, 141–143; waxing and waning of symptoms, 138
muscular dystrophy, congenital form, 110
Muscular Dystrophy Association, 113

NARIC, information retrieval service, 172
National Center for Health Statistics, census data on disabilities, ix
National Council of Independent Living, 241
National Council on the Handicapped, 241
National Federation of the Blind, 234
National Spinal Cord Injury Association, 234
neonatal complications, and parent-infant bonding, 91
neurofibromatosis, 245

neurological damage, cognitive and behavioral impairments due to, 55
neuropsychological assessment: attentional deficits, confused with cognitive deficits, 67; historical background, 58–61; insensitive versus sensitive tests, 59–60; Luria's method, 73; of patients with emotional and motivational limitations, 65–75; of patients with hearing impairment, 64–65; of patients with motoric slowness, 62; of patients with physical limitations, 61–65; of patients with both physical and emotional problems, 60; of patients with sensory or motor disabilities, 63; of patients with visual impairments associated with cerebral lesions, 64; of visually impaired patients, 63–64; strategies, 71–74. *See also specific tests*
Nolan, Christopher, x-xi

organic brain dysfunction, impact on development of children with cerebral palsy and spina bifida, 89
organic hyperkinetic disorder, use of medications for, 100
organicity, a unitary concept, 58

parenting the handicapped child, 91–94; education, 98; family support groups, 99; role, 98; sexual counseling, 100; social encounters outside home, encouragement of, 100
personality development, Eriksonian stages in children with cerebral palsy and spina bifida, 88
physiognomy, "science" of, 152
Porteus Maze Test, 59
Primary Mental Abilities Test, 63
proprium (self-concept), 154
psychosocial assessment: methods, 51–53; primary, not secondary task, 43; purpose, 49–50; sources of data, 47–48; types of data, 48–49; written report, 53
psychosocial impact of disability: acting-out behavior due to communication difficulties, 96; acting out due to immobilization, 96; confusion about or low expectations of capabilities, 96; egocentric personality development due to greater attention from adults, 96; identity issues, 96; low self-esteem, 95; on peers and society, 94–95; personal and emotional problems, 95–96; physical access issues, 95; poor social skills, 95; rebellion against authority, 96; sexual identity issues, 96; severe depression, rare, 97; social withdrawal and isolation, 97
psychosomatic ailments, 9

Queen, Richard, 142

Raven's Progressive Matrices, 64, 70
rehabilitation, defined, 170
Rehabilitation Act of 1973 (with Section 504), 224, 235
Rehabilitation Services Administration, 237
Rey Auditory Learning Task, 71

*Schadenfreude*, 149
scotoma, 64
Section 504, Rehabilitation Act of 1973, 224
self: body as, 7; definitions of, 17–18; meaning structure, plasticity of, 25; and system of personal integration, 17–18, tasks, functions, processes of, 18
self-image, assistive devices as part of, 170, 179
sensory and perceptual deficits, impact

sensory . . .deficits (*continued*)
   on development of children with cerebral palsy and spina bifida, 89
separation-individuation, in children with congenital or perinatal disability, 88
sexuality of disabled persons: abuse, higher potential for, 188–189; acquired versus traumatic disability, 189–191; adjustment, predictors, 189; assertiveness training, 189; attitudes toward, changes in, 183–184; *Coming Home,* movie, 184; congenital disability, 185–189; couples counseling, 192; dependency issues, 185; diversity of personal experience, population, needs, 183–184; familial reactions to, 185; fear of rejection, 193; group therapy, 186; individual psychotherapy 186–187; issues of partners, 191–193, 197; lack of information provided on, 185; Masters and Johnson, intervention techniques, 190; masturbation, 189; peer counseling, 198; persons with spinal cord injury 190; sensate focus exercises with spinal cord injuries, 190; sex therapist, qualities needed, 197–198; sexual surrogates, 187; social skills issues, 186, 193–197
sexual surrogates, difference between prostitutes and, 187
sit-ins for disabled persons' rights, 235
social Darwinists and disability, 252
social learning model of rehabilitation versus medical model, 130–131
social perceptions of disabled persons as: dependent, 248; different, 249; disposable, 249; evoking compassion, 249; identified with their disability, 247–248; mentally incompetent, 248
Spatial Relations subtest (of Primary Mental Abilities Test), 63
spina bifida, incidence, 81. *See also* cerebral palsy and spina bifida

spinal cord injury patients: changes in lifestyle after injury, 118; characteristics of, 117, 127–129; cultural issues and, 132; denial in, 121, 123, depression, lack of, 124–125, 133; depression, treatment for, 133; distress, overreporting of, 126; educational or social learning model in rehabilitation, 120; incidence of, 83, 117; learned helplessness in, 129–130; locus of control in, 129–130; management strategies of, 131–133; medical model of rehabilitation, 120–121, 130–131; motivation and adjustment, 129–130; noncompliance, 131–132; psychosocial reactions to, 121–123, 127, 138; psychosocial issues for, 118; rehabilitation process of, 119–121; self-destructiveness, lack of in, 118; sensory deprivation and, 122; sports and physical therapy for, 133; stage theory of adjustment, 121, 123–127; training in adaptive daily living, self-care skills, mobility techniques, 133; training in interpersonal and community living skills, 133; use of mediator, 132
"spoiled identity," in children with cerebral palsy and spina bifida, 89
stage theory of adjustment to disability, 121, 123–127
stigma: and facial disfigurement, 147, 150; in children with cerebral palsy and spina bifida, 89
stress: defined, 150–151, 177; role in multiple sclerosis, 141–143
stroke, patient re-experiences sense of body as a result of, 5
Strong Vocational Interest Blank, 128–129

Tactual Performance Test, 59, 61, 69
technology, defined, 176; use to increase adaptation of disabled persons, 175
Token Test, 60

# INDEX

Trailmaking Test, 61, 62, 64, 74

Vocational Rehabilitation Act of 1973, Section 504, 222

Wechsler Adult Intelligence Scale: Arithmetic subtest, 73; Digit Symbol subtest, 60; Picture Arrangement, 63, 64

Wechsler Intelligence Scale for Children–Revised, deaf norms, 65

Wechsler Memory Scale, 68, 71

Wide Range Achievement Test, large-print version, 63

workers' compensation: and the courts, 218; disputes about, 201–202; goals, 202; laws, 202; no-fault, 202–203

World Institute on Disability, 241